IITK Directions

Volume 1

Series editor

Amalendu Chandra, Indian Institute of Technology Kanpur, Kanpur, Uttar Pradesh, India

IITK Directions is an institutional series of the Indian Institute of Technology Kanpur. IITK Directions aims at presenting original contributions and reviews of important and cutting-edge topics by the faculty members and other stakeholders of IIT Kanpur in a consolidated manner. It is a platform that reflects upon the ongoing research and development activities, major achievements, and evolving trends. Each volume in the series focuses on a particular area of science and technology with chapters written by the faculty members, students and research staff. Each chapter is written in a journalistic tone for the peer group—readability and accessibility being important parameters. The volumes contain a survey of the subject as a whole and the extent of contributions recorded by the Institute faculty and students who participate so extensively in research.
The volumes in this series will be useful to researchers, practitioners, students, industry leaders, government officials and policy makers. Above all, the series serves as platform to publish the most cutting edge research summaries and evaluate the applications and impact of the research work undertaken at IIT Kanpur.

More information about this series at http://www.springer.com/series/15345

B.V. Phani · Sameer Khandekar
Editors

Innovation, Incubation and Entrepreneurship

Case Studies from IIT Kanpur

Springer

Editors
B.V. Phani
Department of Industrial and Management Engineering
Indian Institute of Technology Kanpur
Kanpur, Uttar Pradesh
India

Sameer Khandekar
Department of Mechanical Engineering
Indian Institute of Technology Kanpur
Kanpur, Uttar Pradesh
India

ISSN 2509-6591 ISSN 2509-6605 (electronic)
IITK Directions
ISBN 978-981-10-3333-9 ISBN 978-981-10-3334-6 (eBook)
DOI 10.1007/978-981-10-3334-6

Library of Congress Control Number: 2016960771

Printed on acid-free paper

This Springer imprint is published by Springer Nature
The registered company is Springer Nature Singapore Pte Ltd.
The registered company address is: 152 Beach Road, #21-01/04 Gateway East, Singapore 189721, Singapore

Series Editor's Preface

I am extremely happy about this special issue of Directions which has focused on the innovation, incubation, and entrepreneurship activities at IIT Kanpur.

With the significant growth of intellectual property at IIT Kanpur along with the modernization of various central facilities, there is a new mandate of integrating these facilities with the new endeavors which will initiate a sustainable growth of our innovation ecosystem. Today, the extent of this ecosystem is rapidly progressing with its activities spanning over Rural Technology, New Product Development, and Entrepreneurship in one hand and development of the futuristic manufacturing technologies in the other. From this point of view, we have chosen the articles which give us a broad framework of interconnectivity between various subsystems. The articles from the Central Workshop and 4-i laboratory have described the futuristic manufacturing potential along with changes in the organization structure initiated to facilitate these changes. Similarly, various articles from start-up cells and product development experiences may give one a unique exposure to the efforts made for sustainable development of the innovation ecosystem. I am thankful to the authors for their spontaneous participation to contribute these articles and make timely revisions based on the editors' notes. I am also thankful to the editors and the publication team for raising the quality of the publication.

It is worthwhile to note that this is the first issue of Directions which is published by Springer. This may bring two opportunities for us: Firstly, we expect to reach a wider community through Springer's platform and secondly, the production quality of the articles is expected to be at a much higher level as every article has undergone a two-stage editing process.

Prof. Amalendu Chandra
Dean, Research and Development
Indian Institute of Technology Kanpur
Kanpur, India

Preface

Post-independent India has come a long way from a 'developing, third-world poor nation' to a vibrant middle-income, emerging market, consumerist economy-driven democracy with an enviable young workforce. We are at a crossroad today with a clear demographic advantage on our side and an opportunity of unprecedented dimensions knocking at our door steps. While business and entrepreneurship skills reside in our psyche from time immemorial, their union with modern science and technology is what will catalyze a paradigm shift in the existing levels of unemployment taking the country forward to newer heights. High-quality technical education on one hand and development of technology-based entrepreneurship skills on the other, working in unison, are the only potent solutions for improving the quality of life and generating employment in our country. As a community, IIT Kanpur realized the importance of this mission soon after our country's economy was opening up in the mid-nineties. Being a premier and leading technology institute in India, we were one of the pioneers in establishing a technology-based business incubator within our campus.

Today, the SIDBI innovation and Incubation Center (SIIC) nurtures a vibrant ecosystem, operating under the Office of the Dean of Research and Development. Our mission is to become an exemplary unit for technology-based mentorship, innovation, intellectual property management, and entrepreneurship, eventually leading to definitive and inclusive societal impact.

It is our pleasure to present before the readers a bouquet of some pertinent case studies and our shared experiences in various facets of our activities. We hope that this issue of *Directions* will be useful for prospective business incubators, managers of technical schools, groups aspiring to start tinkering laboratories, start-up accelerators, and all stakeholders in the fast emerging innovation, incubation, and entrepreneurship space in the country.

We, as editors of this issue, have thoroughly enjoyed our work in compiling this issue, which, incidentally, is also the first edition of *Directions* being published by Springer. All the contributors have done a splendid job in bringing forward the nuances of their activities, which eventually contribute to the strong innovation and entrepreneurship ecosystem of the institute. We must mention here that several new initiatives are under active planning as well as execution. This includes the construction of Science and Technology Research Park, finalizing the Student Entrepreneurship Policy, execution of the INVENT Social Entrepreneurship Program, expansion of the Tinkering Laboratory, establishment of the NIDI-Prayas Center, and implementation of Entrepreneurship In Residence program, to name a few. There is a long way to go and indeed best times are yet to come! We hope that the readers from different cross sections will not only enjoy the articles, but also get inspired to join hands in different capacities to develop the vital ecosystem in the country.

Kanpur, India

B.V. Phani
Sameer Khandekar

Project Team

Amalendu Chandra
Dean, Research and Development
Indian Institute of Technology Kanpur

B.V. Phani
Department of Industrial and Management Engineering
Indian Institute of Technology Kanpur

Sameer Khandekar
Department of Mechanical Engineering
Indian Institute of Technology Kanpur

Reema Mittal
Editorial Consultant

Chitralekha Bhattacharya
DORD Office
Indian Institute of Technology Kanpur

Acknowledgements

The Office of the Dean of Research and Development and SIDBI Innovation and Incubation Center gratefully acknowledges all the funding agencies, support groups, well-wishers, alumni and administrative functionaries of the institute for their continuous support and encouragement, without which the center would not have achieved success. We have always received tremendous backing from the different ministries of the Central Government, Government of Uttar Pradesh, SIDBI, TiE UP, Kanpur Angels, NGOs working with us, MSME officials, Patent Office, and other stakeholders of the ecosystem. The editors acknowledge the active support from Mrs. Sudha Selvaraj, Manager, SIIC, and other staff members of the center. Administrative support from Mrs. Chitralekha Bhattacharya and Mrs. Reema Mittal, Editorial Consultant for this project, is also acknowledged. We acknowledge the support from Student Gymkhana and E-Cell of IIT Kanpur. Last but not the least, we thank all budding entrepreneurs, past and present, who have taken SIIC to newer heights and in the process enriched the institute. Needless to say, we welcome all future entrepreneurs to join this exciting journey.

Contents

About the Editors

Dr. B.V. Phani is currently a Professor of Finance, Innovation and Entrepreneurship at IIT Kanpur. He was earlier the Associate Dean Innovation and Incubation and handled the additional responsibility of a Coordinator, SIDBI Innovation Incubation Center (SIIC) for a decade. Currently, he is the Dean Resources and Alumni with additional responsibility of a Faculty In-charge, Motwani Incubator & Accelerator (MIA) and Coordinator, Syndicate Bank Entrepreneurship Research and Training Center (SBERTC). He is a mechanical engineer with a Ph.D. in finance from IIM Calcutta. Prior to pursuing a Ph.D. at IIM Calcutta, he held various managerial positions including his last stint as the CEO of BCIT Limited—a computer education and services start-up company. He is also the recipient of various teaching and research awards from national and international foundations. He often shares his experiences at various universities across the world as an invited guest. On the entrepreneurship front, the incubation center at IIT Kanpur has undergone several developmental changes under his able stewardship. Currently, the center has 32 start-up firms in diverse areas and has graduated around 30 firms successfully with a few exits. Dr. Phani was also the NEN faculty resource for their 'Educating the Educators (ETE)' initiative across the country and continues to serve as an advisory board member of NEN. Dr. Phani has also been instrumental in taking IIT Kanpur's IP initiative efforts to the next level by facilitating the patenting of a few hundred patents in the last few years from IP generated as part of the institute's research effort. His team has been able to successfully commercialize around 25% of the patents generating a substantial revenue stream and also jump starting IITK's industrial collaboration initiative. He has helped widen the platform for start-up prospective entrepreneurs and intrapreneurs to convert their innovative ideas into commercially viable products. He is also a mentor director in around 25 start-up firms. He is a member of the mentor board of Ivy Cap Ventures and many other such venture, angel, and professional bodies.

Prof. Sameer Khandekar is affiliated to the Department of Mechanical Engineering, Indian Institute of Technology Kanpur, India, since September 2004. He completed master's degree from IIT Kanpur, India (1998–2000), in fluid-thermal engineering and subsequently earned doctoral degree from University of Stuttgart, Germany (2000–2004). Earlier, after his undergraduate studies in Mechanical Engineering from Government Engineering College, Jabalpur (MP), he worked as a marine power plant engineer on board sea-going merchant vessels for four years (1994–1998). He is a recipient of P.K. Kelkar Research Fellowship (IIT Kanpur; 2008–2011), DAAD Fellowship (2011), Prof. K.N. Seetharamu Award (Indian Society of Heat and Mass Transfer, 2010), George Grover Medal (International Heat Pipe Committee, 2007), and Young Scientist Award (Department of Atomic Energy, India, 2005). He is a member of the International Heat Pipe Committee and editorial board member of two international journals on science and technology of heat pipes. He has served as an academic senate member of one central and one state funded autonomous engineering institutes. He has also served as an invited faculty member at four international universities in Germany, France, Brazil, and Thailand. He has over forty-five research publications in international journals, over sixty publications/presentations in international conferences, including 15 Keynote lectures/Invited Talks and two books to his credit. He is presently serving as the Associate Dean Innovation and Incubation of IIT Kanpur (2015–ongoing). His current research interests are in experimental microscale phase-change thermo-fluidic systems, boiling and condensation, heat pipes, and energy systems.

Bridging the Technology Gap: Linking IITs with ITIs

1

S.C. Srivastava

Abstract
Indian Institute of Technology Kanpur has set up an incubation facility in the area of power generation, distribution, transmission, wiring and electrical equipment, with the support from the Ministry of Skill Development and Entrepreneurship (formerly called as the Ministry of Labour and Employment). Primarily focused on bridging the technology gap between IITs and ITIs, the program intends to train a selected number of students in various aspects of setting up and running successful businesses in the power sector.

Keywords
Technology gap • Skill development • Grassroots innovations

1.1 Introduction

Currently, a three-tier system of technical education prevails in India. First tier comprises of engineering colleges and technical institutes/universities offering undergraduate and postgraduate degree courses; second tier includes polytechnic colleges offering diplomas in engineering courses; and third tier covers Industrial Training Institutes (ITIs) offering 1–3 year vocational training programs in 73 engineering and 48 non-engineering trades. While the first two tiers fall under the Ministry of Human Resource Development (MHRD) and All India Council of Technical Education (AICTE), the third tier is governed by the Directorate General of Employment & Training (DGE&T) under the Ministry of Labour and Employment (MoLE), now moved under the Ministry of Skill Development and Entrepreneurship (MoSDE), Government of India.

As per the data available for the year 2013–2014, a large number of students opt for polytechnics and ITIs (Table 1.1). Despite this huge number, most of the efforts in the past have been targeted toward modernizing and enhancing the engineering undergraduate and postgraduate degree education system in the country leaving behind the other two tiers unattended. Today, the biggest challenge is to effectively train this huge pool of young resources in skills that enhance their productivity thus paving way for inclusive growth in the country.

The Government of India has adopted a National Policy for Skill Development and Entrepreneurship 2015 with the primary objective of "*meeting the challenge of skilling at scale with speed, standard (quality), and sustainability.*" It clearly focuses on promoting skill development at ITI level through grassroots innovation, linking them to the nodal entrepreneurship centers in the country and promoting entrepreneurial skills in them.

1.2 MoLE Center at IITK

Recognizing the commendable contribution made toward promoting the entrepreneurial ecosystem in the country, Indian Institute of Technology Kanpur has been chosen as one of the institutional frameworks to carry out the *Skill India* mission effectively. The DGE&T under MoLE/MoSDE has sanctioned funds to the institute to set up an incubation and training facility in the campus. The institute will primarily be responsible for introducing a selected number of ITI graduates to the concept of

S.C. Srivastava (✉)
Department of Electrical Engineering, IIT Kanpur, Kanpur, India
e-mail: scs@iitk.ac.in

B.V. Phani and S. Khandekar (eds.), *Innovation, Incubation and Entrepreneurship*,
IITK Directions 1, DOI 10.1007/978-981-10-3334-6_1

Table 1.1 Technical education in India—Statistics (2013–2014)

Technical education program	No. of colleges	Intake each year
Undergraduate engineering & technology[a]	3,384 (more than 4,100 now)	16,34,596
Postgraduate engineering & technology[a]	2,132	63,430
Diploma engineering & technology[a]	3,436	11,35,179
Industrial Training Institutes (ITIs)[b]	11,964 (2,284 Govt., +9,680 Pvt.)	16,92,836 (Seating Capacity)

[a]*Source*—AICTE Website; [b]*Source*—MOLE/DGE&T Website

entrepreneurship and training them in the various aspects of setting up and running successful businesses in the power sector. The institute has partnered with Regional Advanced Training Institutes and the World Bank for operationalizing the incubator.

1.3 Objectives

The primary objectives behind this initiative are as follows:

- To set up an incubator in the area of power generation, transmission, distribution, wring and electrical equipment. The electricity sector over past two decades has witnessed tremendous growth thereby marking the per capita electricity consumption as an important indicator of the economic growth of a country. Currently, India's per capita electricity consumption stands at 1000 kWh, which is much less than that of the few developed countries, ranging between 10,000 and 25,000 kWh, and even below the world average of about 2500 kWh. The sector is growing at a fast pace, and new technological changes in the generation, transmission, and distribution of electricity as well as at the consumer end are taking place. Therefore, there is a critical need of innovative solutions in this sector.
- To train selected ITI students/graduates in all major areas of business operation and to
  - facilitate their skill development,
  - expose them to new technologies in the sector,
  - assist them in identifying appropriate technology,
  - provide them hands-on experience on working projects,
  - help them in project/product selection, and
  - guide them in business opportunities and commercial aspects.
- To allow ITI students/alumni to access the existing SIIC incubation facilities for incubating their ideas or assisting the existing incubates.

1.4 Planning and Infrastructure

The MoLE Incubation and Training laboratory is being set up in the southeastern block of the institute's workshop building where currently Tinkering laboratory is located (Fig. 1.1). Since all the necessary equipments have already been purchased for the above facility, a temporary laboratory facility has been set up in the old Student Activity Center (SAC), Room 102 (Figs. 1.2 and 1.3); the machining equipments, however, have been placed in the workshop building.

A management committee has been formed to periodically review and provide guidance to the center. The process has already been initiated with the Regional Directorate of Apprenticeship Training (RDAT) and the Advanced Training Institute (ATI) Kanpur to select the first batch of ITI graduates in the program. A maximum of 30 students will be taken every year and at any point of time. Each trainee will be given sustenance allowance. They will undergo 3 months of training on the development of core, business, and entrepreneur skills. During this period, each trainee will develop a project proposal and submit it to the managing committee for review. The commercially viable proposals will be recommended for incubation at the SIIC center for a

Fig. 1.1 Site for the MoLE/MoSDE laboratory

Fig. 1.2 Energy meter test bench in old SAC

Fig. 1.3 Programmable logic controller (PLC) trainer

further period of maximum 9 months. The facilities will also be used for imparting regular training to ITI/diploma students.

1.5 A Way Forward

The MoLE facility has recently commenced its operations in the summer of 2016. The first batch of 30 students from the technical institutes in the region will be given training in relevant electrical and mechanical engineering aspects besides being trained in entrepreneurship, legal and accounting procedures, IP management, business development, and marketing. It is only after a year that one would be able to gauge the success of this initiative. However, the program intends to focus on providing latest hands-on training to students and instill the entrepreneurial spirit in them.

Innovation and Entrepreneurship Ecosystem at IIT Kanpur: A Journey of Serendipity

2

Sameer Khandekar and B.V. Phani

Abstract

The journey of an entrepreneurial firm towards success is fraught with landmines of failure interspersed with serendipitous opportunities. Building and sustaining an ecosystem to nurture these efforts to fruition, particularly in an academic setting, is not only humungous, but also faced with insurmountable challenges at every turn. There is no pre-defined path or an established formula in achieving this objective. Each ecosystem has to traverse its own path constantly learning from its own experiences and evolve into a viable and sustainable ecosystem. This article traces this journey at IIT Kanpur over the past one decade providing rare glimpses into the domain of building and sustaining an innovation and entrepreneurship ecosystem anchored to an academic institute or a technical institute of higher learning within the framework of a national innovation ecosystem.

Keywords

Academic institution • Innovation ecosystem • Entrepreneurship • Technology incubator

2.1 Introduction

The seed of the innovation ecosystem at Indian Institute of Technology Kanpur was sown in late 1990s with the active support and engagement of Small Industries Development Bank of India (SIDBI) as a natural extension to its existing efforts of being the banker for small and medium enterprises of India. This seed was further nurtured by the Department of Science and Technology (DST), which was assigned the mandate of creating entrepreneurial ecosystems at the national level in collaboration with institutions of learning. DST over the last decade and a half has been able to nurture more than 100 such islands of vibrant entrepreneurial hubs in academic settings across the length and breadth of the country. This effort by DST has been further augmented by government departments, such as Department of Electronics and Information Technology (DEITY), Department of Biotechnology (DBT), Ministry of Small and Medium Enterprises (MSME), Ministry of Labour and Employment (MoLE; now called as the Ministry of Skill Development and Entrepreneurship) and Department of Scientific and Industrial Research (DSIR), to name a few.

These efforts have led to the founding of over 20,000 start-ups and growing, valued at $75 Billion directly employing over 300,000 people. A recent report by NASSCOM puts India in the third place globally, having 4200 new age companies and growing at the rate of three to four start-ups per day with funds in this sector having reached a $5 Billion (Rs. 325 Billion @ 65/$) mark in 2015.

S. Khandekar (✉)
Department of Mechanical Engineering, SIDBI Innovation and Incubation Center, Indian Institute of Technology Kanpur, Kanpur, 208106, India
e-mail: samkhan@iitk.ac.in

B.V. Phani
Department of Industrial Management and Engineering, Indian Institute of Technology Kanpur, Kanpur, 208106, India
e-mail: bvphani@iitk.ac.in

B.V. Phani and S. Khandekar (eds.), *Innovation, Incubation and Entrepreneurship*, IITK Directions 1, DOI 10.1007/978-981-10-3334-6_2

2.2 The Genesis

It was a chance visit to Israel in 1993 by Prof. Sanjay G. Dhande, the then Dean of Research and Development, who observed there that each academic institute had an internal entrepreneurial ecosystem channelizing the students creative and innovative ideas into business propositions and enabling them to monetise the same. This idea to offer students an alternative career option of being job creators instead of job seekers, triggered the imagination of Dr. Dhande who believed that it was time for the highly subsidized elite institutions like IITs to nurture job creators instead of job seekers, thus enabling these institutions to contribute to the economy by leveraging their intellectual wealth.

A chance call from SIDBI, Lucknow, inviting a proposal to be submitted at a short notice, for setting up an innovation centre at the institute led to the founding of what is called today as the SIDBI Innovation and Incubation Centre (SIIC) in the year 2001. The centre eventually started its operation in 2004. This article would be incomplete without acknowledging the contribution of Prof. Sanjay G. Dhande and Prof. Ashok Mittal for their efforts in making this possible, and Dr. Rahul Varman's role as the first coordinator to operationalize the same. Dr. Dhande, who went on to become the Director of the institute and held the position for 10 years, was a pillar of strength and support; not to mention Prof. Ashok Mittal's contribution in nurturing this ecosystem, transforming the single incubation centre having a 10,000 sq. ft. footprint into a vibrant innovation ecosystem having a 100,000 sq. ft. footprint (and still growing!).

SIIC was the first small step towards transforming IIT Kanpur into an innovation hub. It was way ahead of its time, as the external ecosystem was not ready and finding entrepreneurially minded people to take advantage of the system were few and far in between. Fostering innovation, research and entrepreneurial activities in technology-related areas, it is primarily focussed on:

- Creating a generation of zealous entrepreneurs.
- Converting novel research into valuable intellectual property.

Since the beginning, the centre was made accessible to everyone. In this, we believe SIIC is the only incubation centre anchored to an academic institution which opens its doors to any entrepreneur from anywhere in the world without any reservations. Any limited liability company promoted by any individual desiring to nurture their idea could contact the centre, seeking mentorship and incubation. However, generating trust amongst start-ups and convincing IITK students and faculty to come forward and turn their innovative ideas into profitable ventures turned out to be a humongous task. The fully air-conditioned incubator capable of housing 16 ventures at that time laid vacant for almost a couple of years before entrepreneurial ventures started trickling in. (Today, SIIC boasts of over 30 incubates thriving in the ecosystem.)

2.3 The Learning Curve

As SIIC offered its services to entrepreneurial ventures registered with it, it realized that it still needed to learn the ropes of the trade before gaining a strong foothold as a technology incubator. A major learning was that, for successful incubation, the incubator has to leverage the strengths of its host institution, which, in this case, is IIT Kanpur. Given that IITK has state of the art laboratories and experimental spaces which are augmented by a large pool of captive intellectual resources such as students, faculty and researchers, opening these resources to be tapped by an entrepreneurial firm would not only decrease the cost of taking the idea to market, both in terms of time and financial resources, but also improve the probability of success of an entrepreneurial venture.

Given the above, SIIC has expanded its services to incubate companies by providing:

1. **Access to IITK Infrastructure**: Till 2001, IITK laboratories were accessible only to faculty and students. To make them available to start-ups as well needed a proper policy framework. The most difficult task was to convince incubate companies to nominate 3% of their company's shares in favour of IIT Kanpur and 1% in favour of the mentor. It was made compulsory for the start-ups to select one faculty member from the institute who shall act as a mentor and guide the company on product development in return of 1% of share equity. However, they can also opt for an industry mentor over a faculty mentor from the SIIC database. After countless meetings and heated debates, the policy framework was put in place. Today, SIIC charges the incubate companies for infrastructure and services, seed loan and IIT Kanpur Intellectual Property in form of service charges and equity share.
2. **Unmatched Facilities and Services**: Besides offering a fully furnished air-conditioned incubation space with all the necessary amenities, SIIC also offers a common pool of soft and hard infrastructure to be shared by all incubate companies. *Moreover, SIIC is the only incubator in the country which provides on-campus furnished residential accommodation to the promoters of incubate companies.* The support services given to the start-ups include mentoring and advisory services as well. In addition to the IITK's expert faculty, specialized mentors are also made available to the companies to either assist with

particular strategic areas or to provide project-oriented consultation.

3. **Numerous Funding Options**: Over a period of time, several funding options have been made available to the incubated companies. Primary funding options include PRISM (Promoting Innovations in Individuals, Start-ups and MSMEs), MSME (Ministry of Micro, Small and Medium Enterprises) and DIT (Department of Information and Technology) support. SIIC further assists the incubated companies in availing more funds from banks such as SIDBI Risk Capital/Venture fund, government funding such as TDB, BIRAC, SIDBI-TIFAC, Strategic Stake from Industry and others. The centre also facilitates funding from venture capitalists and angel investors.

From the recovery of the seed fund and earnings through equity dilution, SIIC has now started its own modest seed fund for disbursement. It has undertaken seed fund disbursement (in the form of soft loan and/or equity option) to more than 20 start-ups through external available funding and facilitated risk capital and venture funding (Stage II and III) to over 10 start-ups amounting to about Rs. 100 crores. IIT Kanpur's equity in top 10 start-ups is valued to be more than Rs. 15 crores.

2.4 Nurturing the Environment

As the centre started progressing gradually, both as an incubator and a technology transfer office, it was envisaged that for it to grow and expand into an ecosystem, a set of activities would have to be undertaken in parallel creating a pipeline of future entrepreneurs.

2.4.1 Developing R&D Infrastructure

Though the IITK laboratories were made accessible to the incubate companies, a need was forecasted to have independent laboratory infrastructure where innovators could experiment with their ideas and develop them into prototypes. Not only this, but the lament of academicians that the students spend their time more on social media and other undesirable pursuits could be effectively addressed by creating a tinkering space with state-of-the-art facilities open round the clock to encourage and harness the creative energies of these future technocrats and igniting the dormant spirit of entrepreneurship. This lead to the setting up of first of its kind laboratory called Tinkering Laboratories in India funded by DST and augmented by alumni contributions. This model of Tinkering laboratories as a creative space has been adopted by the DST and floated across the country vindicating SIIC's foresightedness in this regard.

Extensive infrastructure spread over 10,000 m^2 has now been developed over the years to provide an in-house state-of-the-art support system to all incubates and students, thus meeting their software as well as hardware manufacturing needs within the campus (Fig. 2.1). The incubates and students can avail (a) design and development software, operating systems platforms, computational tools, data storage, etc. and (b) hardware manufacturing needs such as prototype development, precision machining, fabrication, electronics and printed circuit board development, micro-/precise fabrication and welding under one umbrella called as the **Imagineering Laboratory** (Fig. 2.2). This laboratory has three major components (a) Tinkering laboratory (b) 4i laboratory and (c) Motwani ideation accelerator.

Tinkering Laboratory: Started in late 2012, the Tinkering laboratory works in conjunction with the 4i laboratory. It is a platform for creative minds to come out of their 'Think Space' to hands-on 'Tinker Space' in order to transform their ideas into real-time engineering objects and eventually to products and patents. The laboratory is accessible to incubators from all across the country as well as to students and start-up entrepreneurs to test, validate and prototype their ideas into new products and services, specifically in the areas of mechanical/electrical/electronic elements fabrication processes. Tinkering laboratory witnesses at least 18–20 footfalls per day with students from different departments and programs. The facilities available at the Tinkering laboratory include basic lathe and milling machines, 3-D scanner, bench drilling machines, vacuum plastic forming machine, injection moulding machine, sheet metal cutting and bending machine, shearing machine, small grinders, buffing tools, hand grinders and drill machines, metal fitting and wood work activities, spray painting, fitting and carpentry tools, marking and measurement tools, welding machine and air-compressors.

4i Laboratory: The 4i Laboratory, incepted in the year 2003, is aimed at providing high-end manufacturing support to UG and PG activities across the institute and also to help incubate new companies at IIT Kanpur. The laboratory houses machining centres which are mainly classified as CNC centres, non-traditional machining centres, PCB fabrication line and associated equipment, conventional machining centres, etc. All high-end and precision manufacturing jobs are handled by this laboratory.

Motwani Ideation Accelerator: Established with the generous funding by Motwani Foundation, Motwani Ideation Accelerator (MIA) is a soft facility in the domain of embedded systems, platforms, algorithms and IT related ideas. The sole purpose of MIA is to facilitate and nurture groundbreaking and innovative ideas primarily sourced from B.Tech. and M.Tech. projects, and Ph.D. dissertations. It is also in the process of linking up with other prominent accelerators in India, such as Indian Angel Network

Fig. 2.1 A snapshot of various facilities under SIIC

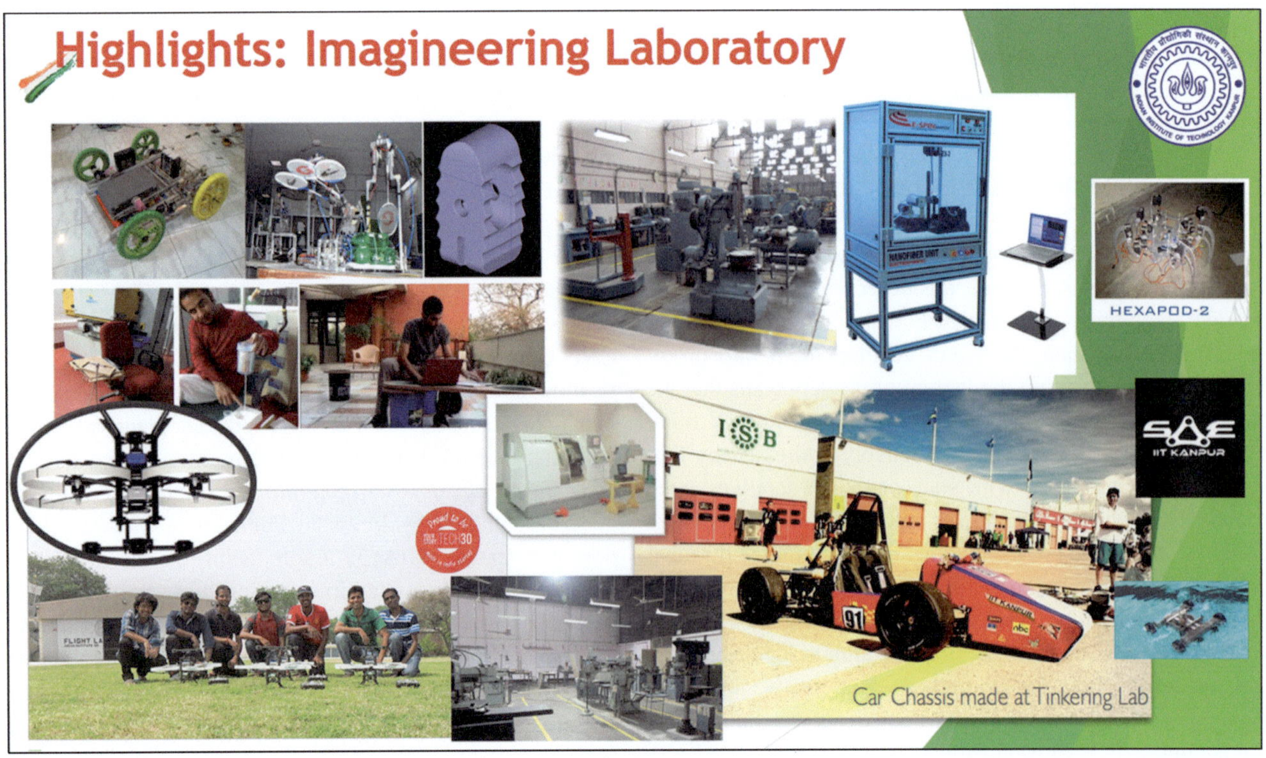

Fig. 2.2 A collage of activities and infrastructure of the Imagineering Laboratory

Incubator, Microsoft Accelerator, Khosla Laboratories, The Hatch, Villgro, and The Startup Village so that students not only gain wider exposure, but also refine their ideas through summer and winter internships at these accelerators. SIIC is in the process of creating two venture funds as a part of this accelerator:

1. *Rajeev Motwani Impact Fund*—It is meant to support groundbreaking ideas specifically that target to benefit the bottom half of the economic pyramid, for e.g. start-ups dealing with poverty alleviation, non-formal learning/education/technology for the poor, mobile health support.
2. *Rajeev Motwani Venture Fund*—It is meant to encourage ideas emerging from Indian academia to help the creation of the next generation mobile internet and health technologies.

Bio-Incubator: Established by BIRAC, under BISS (Bio-Incubator Support Scheme) in 2016, Bio-Incubator is a boon to biotechnology based start-ups as it is situated in one of the India's most prestigious institutions—IIT Kanpur. The expertise and existing resources available at IIT Kanpur enables it to stand out amongst other bio-incubators. As an integral part of innovation ecosystem of IITK, bio-incubator is emerging into other related domains of sciences and is in the process of developing an in-house research park. Besides addressing the needs of bio-innovation and promoting entrepreneurship, bio-incubator at SIIC is adopting advance technologies and working rigorously to remain at the forefront of global health and business world. So far, it has incubated more than 13 companies and successfully graduated 3 of them. The incubator also helps incubate companies to stay abreast with the latest research and upcoming technologies.

MoLE Incubator: This Incubator has been set up to attract the grassroot innovators, typically with Diploma/ITI background, from the technical Institutes in the region working in the area of Power Generation, Distribution, Transmission, Wiring and Electrical Equipment. Apart from technical skill upgradation in electrical and relevant mechanical engineering aspects, all participating candidates will be provided training in entrepreneurship, legal and accounting procedures, IP management, business development and marketing. SIIC has partnered with Regional Advanced Training Institutes and the World Bank for operationalizing the incubator.

Incubation Accelerator: The centre is creating an accelerator in the NCR (National Capital Region) region at Sector 62, NOIDA, to facilitate business development and funding. In addition, a small office facility for accelerator services is planned at the Wall Street, New York, USA.

2.4.2 Promoting Entrepreneurship at Student Level

The most crucial step towards creating a vibrant entrepreneurship ecosystem at the institute was to engage the student community. Towards this end, SIIC together with the student community started an international business competition called Megabucks in 2000, which slowly over a period of 9 years evolved into an Entrepreneurship Cell (E Cell). E-cell is a student senate approved body that organizes several events around the year such as E-Summit, TEDx, E-Factor, Techkriti, business competitions, workshops and seminars, boot camps, and 'unconventions' to provide a platform wherein students are gradually introduced to various facets of entrepreneurship. Besides E-cell, the centre is actively engaged in the functioning of the following:

Business Club—The Business Club aims to inform, inspire and encourage students about business and entrepreneurship. The club enhances one's acquaintance with the business world with the help of Focus groups: Strategy group, FINIIT group (finance enthusiasts of IIT Kanpur) and the Newsletter, and by holding regular lectures, workshops, competitions and strategic games, which unveil the real market situations to the students.

Promotion of Work Experience and Research (PoWER)—It is a student body aimed at promoting and coordinating student research and development through industry–academia interaction. It aims to provide students the freedom and opportunities to work upon diverse areas of business, technology and existing industrial and national challenges, as well as pursue their own ideas so as to become innovators and future leaders. It is a step towards creating a sustainable innovation ecosystem.

SIIC has also conceived and proposes to implement a series of activities to engage, ignite and sustain the entrepreneurial spirit of the students:

- Graduate Entrepreneur Fellowship Program (GEF), where a student, if he or she chooses to embark on the entrepreneurial path immediately after graduation, will be awarded a fellowship during the period of their incubation in lieu of equity.
- Introductory 101 course on Entrepreneurship and Innovation.
- Introduction of Master's program in Entrepreneurship and Innovation.
- SPADE (Systematic Promotion and Development of Entrepreneurship) Program: Under this program, several proactive initiatives for interaction between the stakeholders are planned to be executed by dedicated resource persons, as part of the annual calendar of the institute.

2.4.3 Initiating Outreach Programmes

From the very beginning, SIIC's vision was to contribute towards creating a healthy ecosystem for innovation in the country and therefore it entered into strategic alliances with organizations such as NEN (National Entrepreneur Network), IIM Ahmedabad and others. Together with these, the centre organized regular workshops and trainings in institutes across the country to promote entrepreneurship. In addition, it regularly conducted awareness programs, talks, group discussions, seminars and outreach activities, in association with organizations such as TiE-UP, FICCI, ISBA, SIDBI, MSME and other supporting groups.

The Syndicate Bank Entrepreneurship Research and Training Centre (SBERTC) at the institute is another centre which conducts cutting-edge research, teaching and training in entrepreneurship in the Indian context by utilizing the facilities of SIIC and the existing intellectual pool in the institute.

2.5 Evolution of Innovation and Entrepreneurship Ecosystem

Today, the ecosystem at IIT Kanpur is a unique experimental and incubation space fostering innovation and entrepreneurship. SIIC is not a stand-alone facility contributing to the innovation and entrepreneurship ecosystem at IIT Kanpur, but only one component of it. The ecosystem is driven by a set of laboratories and testing facilities encompassing the entire gamut of engineering disciplines from Bioengineering, Mechanical, Electronics and Electrical to IT and ITES, and is open to anyone with an innovative idea. These facilities provide the students, innovators and entrepreneurs, both within the institute and outside, a unique environment to fructify their ideas into viable high technology intensive products, thus helping them in building a successful entrepreneurial venture. SIIC, along with the Innovation Council guiding its operations, is at the epicentre of this ecosystem fostering, facilitating and funding these nascent entrepreneurial ventures. IIT Kanpur faculty plays an important role in this ecosystem by providing knowledge and technical expertise as mentors to make them globally competitive.

2.5.1 Incubation Centres

From 2000 till date, SIIC has grown from one incubation centre representing IIT Kanpur to nine centres representing various arms of Government (Fig. 2.3). It acts as a

- SIDBI Innovation and Incubation Centre (for SIDBI)
- Technology Business Incubator (for DST)
- MSME Incubator (for MSME)
- Technology Incubation and Development of Entrepreneurs (for MIT)
- Technology Entrepreneurship Promotion (for DSIR)
- Bio-Incubator (for DBT)
- MoLE Incubator
- Social Enterprise Incubator (Newest addition).

The activities at these centres have been supported by initial corpus funds supported by IIT Kanpur, Government of Uttar Pradesh and Small Industries Development Bank of India (SIDBI). Subsequently, funds have been received for supporting seed funding and operational expenditures from Technology Business Incubator (supported by NSTEDB), Department of Information Technology (DIT), DBT, Incubation activities under the Ministry of Small and Medium Enterprises (MSME), Incubation activities under Department of Scientific and Industrial Research (DSIR), Incubation activities in power generation, distribution and transmission (supported by Ministry of Labour and Employment, now termed as Ministry of Skill Development and Entrepreneurship), and Social Entrepreneurship INVENT Program of the Department of International Development (Government of UK).

2.5.2 Portfolio

So far, the centre has mentored over 60 start-ups; out of which 35 companies have already graduated and 30 are still under incubation (Fig. 2.4). There have been four exits from the incubator, wherein equity dilution of IIT Kanpur's share has been achieved with dividends. Several acquisitions have also been achieved, some of which are tabulated in Table 2.1. Many of the start-ups are scaling fast on the success ladder. For more details on the present portfolio, please visit www.iitk.ac.in/siic.

Any private limited company desiring to nurture their idea can contact the centre seeking mentorship and incubation. However, the selection procedure is vigorous where an applicant is scrutinized on the technical and business feasibility of the idea.

2.5.3 IP and Technology Transfer

Since the beginning, SIIC has acted as a nodal agency for IP and Technology Transfer for the institute. It provides professional aid to the IITK faculty and students for filing

Fig. 2.3 Structure of SIDBI Innovation and Incubation Center, IIT Kanpur

patents and copyrights. Besides, the centre facilitates the use of IITK IP if desired by a start-up. The terms and conditions for such IP licensing are decided by the institute. So far, over 350 patents have been filed, out of which about 50 patents, worth over Rs. 3.0 crores, have already been successfully commercialized (Fig. 2.5).

SIIC also facilitates the modification and upgradation of the software/products developed by the faculty/students of IIT Kanpur to suit the industry requirements with the help of a commercial partner. The concerned faculty member acts as a mentor. The commercial partner is also responsible for marketing the product and providing customer support.

2.6 The Way Ahead

Since its inception, the centre has grown tremendously and has emerged as a prestigious incubator in India. It won the National Award for Technology Business Incubators for the year 2011. The centre continues to grow, and the future looks bright. In the immediate future, SIIC is undertaking two major initiatives for which funds have already been generated:

Establishment of an Industrial and Translational Research Park. IIT Kanpur has been performing high-end research since its inception. The sponsored research funding

Fig. 2.4 Statistical overview of SIIC, IITK portfolio

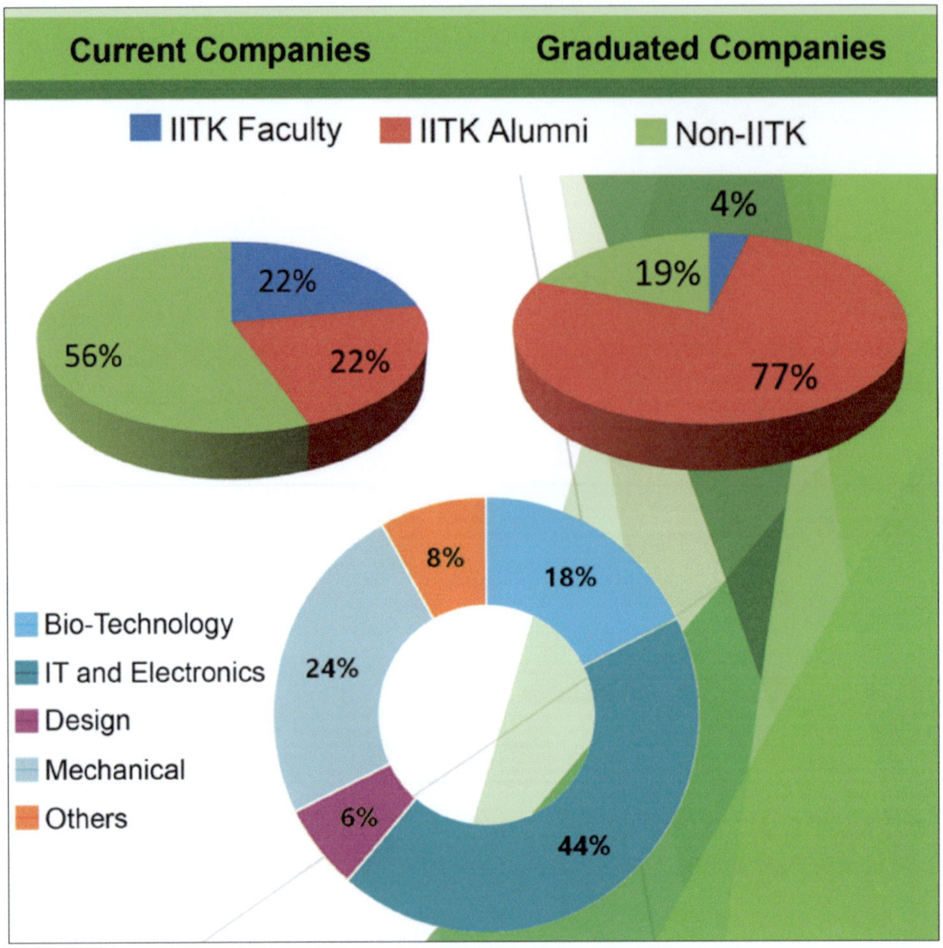

Table 2.1 Mergers and Acquisitions

Incubate company	Strategic stake/merger
Adya Systems	Whiz Net Inc., USA
Aurora Integrated Systems	Tata Group
Whorl Engineering	Walchandnagar Industries
Weather Risk	ICICI Lombard
Aarsh Management	CMC Ltd.
Geokno India	GMR Infrastructure
Irene Nanocore	Biocon

and research output of the institute has grown significantly over the years, and the success in various disciplines has given it national and international visibility. Simultaneously, the innovation and entrepreneurship ecosystem of the institute has seen tremendous growth. The current level of engagement with the industry makes evident the rich ecosystem at IIT Kanpur. The nature of the interactions with the industry has helped in identifying the contributing units of the ecosystem. However, the scale of these interactions needs to be enhanced. This will also facilitate the goals of discovery, knowledge creation, commercially viable patents, spin-off ventures, leading to much needed innovation and entrepreneurship development. This requires new investments for improving the research and development infrastructure on the campus. Perhaps, the most important step required in this direction is the setting-up of a state-of-the-art Industrial and Translational Research Complex. The vision of the proposed complex is to provide easy access to infrastructure, equipment and technical knowledge base to the industry and academia to create a sustainable and affordable industrial research ecosystem in the institute.

Industrial and Translational Research Complex has been conceived as a new kind of research and development space within IIT Kanpur. It will be primarily used for housing

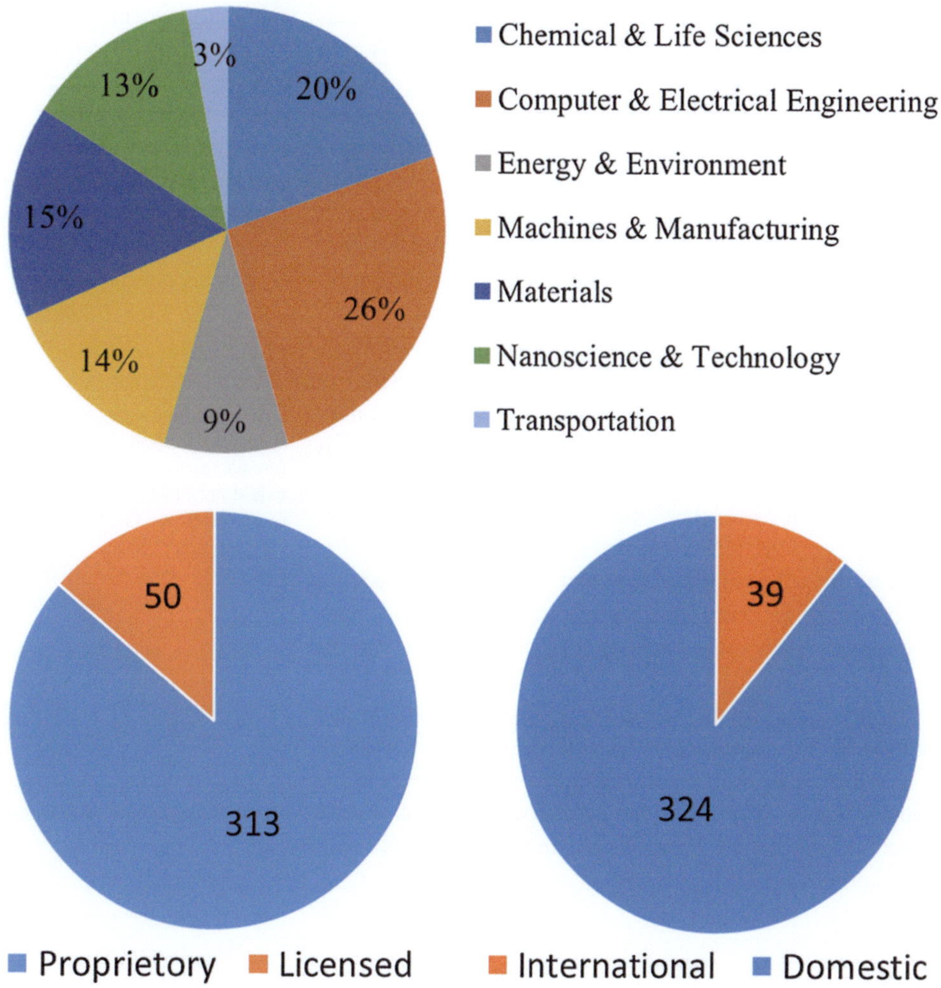

Fig. 2.5 Patents statistics of the institute

industry-sponsored laboratories and incubating technology-oriented companies. The institute will evolve suitable policies, in consultation with the Institute R&D Committee, to allocate and manage space in the Innovation Research Complex for industries research units and their projects thereof. Some part of the Research Complex will be used for housing central research facilities, which can be used by any researcher across the institute. The complex will stimulate innovative research in cross-disciplinary areas that hold promise for significant development in the years ahead. The complex will encourage faculty, students and entrepreneurs/industrial units from diverse disciplines to interact in a mutually supportive manner. This concept of an integrated complex is probably being tried for the first time in an academic institution in India and could become a trailblazer.

Establishment of a Social Incubator. The Technology Development Board (TDB), Government of India in partnership with the Department for International Development (DFID), UK, have initiated the 'Innovative Ventures and Technologies for Development (INVENT)' program. TDB, being the anchor of the program, has chosen Villgro, India's oldest and foremost social enterprise incubator, to execute this program. This program aims to support up to 400 entrepreneurs in the 8 low-income states (UP, MP, Bihar, Chhattisgarh, Jharkhand, Rajasthan, Orissa and West Bengal) and make at least 50 of them investable in the next 5 years. Based on its 15 years of experience in incubating social enterprise, Villgro has decided to support 4 other incubators, and SIIC, IIT Kanpur, is one of them. The INVENT program, to be launched in the summer of 2016, will provide selected incubators with technical and financial assistance by way of mentoring, networks, know-how, templates, etc., to make selected incubators successful at social enterprise incubation. The ultimate goal is to create a viable social enterprise (for-profit) pipeline for impact investments in the above mentioned low-income states (LIS) of India. SIIC is looking forward to participating in this exciting program and bring IIT Kanpur to the forefront of social and impact incubation in the country. If more investments are made available to test new ideas generated by India's entrepreneurs, India would be well placed to develop solutions to entrenched global development challenges, of relevance not only to India, but also to low-income countries.

2.7 Summary and Outlook

India is a country of incredible contrasts—post independent India has come a long way from a 'developing, third-world, poor nation' to a vibrant middle-income, emerging market economy-driven democracy with an enviable young workforce. While business and entrepreneurship skills resides in our psyche from time immemorial, its union with modern science and technology is what will catalyze a paradigm shift in the existing levels of poverty. High-quality technical education on the one hand and development of technology-based entrepreneurship skills on the other working in unison is the only potent solution for improving the quality of life and generating employment in the country. Technology incubators certainly have a definitive role to play in the emerging scenarios.

In this background, IIT Kanpur has created a unique ecosystem for nurturing innovation and entrepreneurship within the academic campus, singular in many aspects in the entire country. While SIIC has managed to achieve several benchmarks, there are many challenging aspirations not only from the student and entrepreneurship community in particular, but also from the nation as a whole. In this exciting journey, SIIC is committed to help create and manage intellectual property for the institute, set up an exemplary paradigm for technology incubation and to catalyze entrepreneurship for positive socio-economic impact. The centre strives to provide even robust platforms in the near future for start-ups and prospective entrepreneurs, including IITK students, faculty and those in the society at large, for converting their innovative ideas into commercially viable products and services.

Acknowledgements SIIC gratefully acknowledges all the funding agencies, support groups, well-wishers, alumni and administrative functionaries for their loyalty and support without which the centre would not have achieved this success. Inputs for this report have been received from Prof. Amalendu Chandra, Dean of Research and Development; Prof. J. Ramkumar, Professor-in-charge, Tinkering Laboratory; Prof. Shantanu Bhattacharya, Coordinator, 4i Laboratory; Prof. S. C. Srivastava, Coordinator, MoLE Laboratory; Prof. Amitabha Bandopadhyaya, Coordinator, Bio-Incubator facility and Prof. Bishakh Bhattacharya, Head, Central Workshop. The authors also acknowledge the active support from Mrs. Sudha Selvaraj, Manager, SIIC and other staff members of SIIC, Office of the Dean Research and Development and Office of the Dean, Resources and Alumni Relations, IIT Kanpur. Last but not the least, we thank all entrepreneurs, past and present, who have taken SIIC to newer heights. Needless to say, we welcome all future entrepreneurs to join the movement.

3 Laboratory to Market: A Case Study

Tarun Gupta

Abstract

A device to collect submicron particles was designed, developed, and field evaluated at IIT Kanpur. This indigenously developed sampler consists of a single impaction stage and an after filter and operates at a flow rate of 10 LPM. The impactor removes particles greater than 1.0 μm from the airstream and uses a backup filter to collect the remaining finer particles. It has a pressure drop of 18.5 cm of H_2O when used with a 47-mm Teflon filter (2 μm pore size). High-vacuum grade silicone grease is used as an impaction substrate. Both wet and dry aerosol generation systems were used for laboratory evaluation of this sampler using an optical scattering-based real-time instrument. PM_1 (particles having aerodynamic diameter (d_a) less than 1 μ) sampler worked well in the field with no sign of particle overloading, wall losses, deformation of impaction substrate, and particle bounce-off within the sampler body. Field testing consisted of aerosol sampling at various outdoor locations within and outside the IIT Kanpur campus. Upon sampling, all the loaded filter substrates were subjected to gravimetric analysis for mass measurements, followed by chemical analysis to quantitatively determine the elemental and ionic concentrations. These data were used for source apportionment using various available mathematical tools/packages. After several years of successful use of this sampler, it was licensed for a 10-year period to a Delhi-based company (Envirotech Instruments Pvt. Ltd.) as a part of academia to industry technology transfer. This PM_1 sampler has recently been (October 1, 2015) launched in the market as APM 577.

Keywords

PM_1 • Gravimetric mass • Chemical composition • Elemental composition • Anions • Technology commercialization

3.1 Introduction

Throughout our lives, we are exposed to environmental contaminants present in air, water, food, and soil. These contaminants can also get transferred from one media to another. Distinct biological, chemical, and geological processes in natural environments can drastically alter the bioavailability and toxicity of contaminants. Frequent occurrence of foul air in our urban cities has become a major public health concern in India. The simultaneous increase in urban population and anthropogenic activities has created extensive potential for pollution-related health damage. Measurement of particle size distribution and chemical composition is necessary in understanding sources and atmospheric processes in the urban airsheds.

Many epidemiological studies have linked particulate matter (PM) and especially the PM2.5 (d_a < 2.5 μm) mass to adverse respiratory and cardiovascular health effects. These health effects include premature mortality, asthma exacerbation, decreased lung function, and increased risk of

T. Gupta (✉)
Department of Civil Engineering, IIT Kanpur, Kanpur, 208016, UP, India
e-mail: tarun@iitk.ac.in

B.V. Phani and S. Khandekar (eds.), *Innovation, Incubation and Entrepreneurship*, IITK Directions 1, DOI 10.1007/978-981-10-3334-6_3

myocardial infraction among others. PM exposure is commonly monitored as mass concentration of PM10 ($d_a < 10$ μm) or PM2.5 ($d_a < 2.5$ μm), although increasing toxicity with decreasing aerodynamic diameter has been reported due to increased surface area and enhanced chemical reactivity with other species.

Kanpur is one of the most polluted cities in India today. Kanpur's climate can be characterized by a very hot and dry summer and very cold winter; temperatures during summers can go up to 48 °C and winters ~0 °C. The major problem for Kanpur city is that a busy highway with both slow and fast-moving vehicles passes through the city. Frequent traffic jams and the exponentially increasing vehicular and population growth rate have led to a very high level of aerosol build-up in the city's air. In addition, contributions from other pollution sources, such as roadside dust, trans-boundary migrations, power plants, solid waste including tannery waste combustion, and local brick kilns, have further deteriorated the air quality. Particulate matter from these sources may contain hazardous pollutants that have carcinogenic and mutagenic effects. Fine (PM2.5) particles are generally emitted from activities such as industrial and residential combustion and from vehicle exhaust. Fine particles particularly submicron ultra-fine particles also called secondary particles are also formed in the atmosphere when gases such as sulfur dioxide, nitrogen oxides, and volatile organic compounds, emitted by combustion activities, are transformed by chemical reactions in the air. PM1.0 fraction has been shown to be most critical and has hardly been studied in detail especially in India.

Unless we adequately understand the amount, transport, and fate of contaminants in the environment, our ability to craft adequate control strategies or regulatory criteria protective of both human and ecological health will be compromised. Hence, at times, accurate measurement of contaminants in various media is considered as the first step toward their effective control. This is followed by the need for the accurate assessment of population exposures that occur in homes, offices, schools, and during transit, as well as in the outdoor environment.

3.2 Major Objectives

(a) Design and fabricate a submicron sampler.
(b) To evaluate its performance in the laboratory as well as in the field.
(c) Transfer this field evaluated PM_1 sampler to an industry which can bring to the market its commercial version.

3.3 Theoretical Design Considerations

A single-stage impactor separates the particles into two size groups. An accelerated jet of aerosol is forced through a nozzle, and it faces an obstruction (impaction substrate) in its path which forces it to make a sharp 90° turn. Particles having sufficient inertia continue to travel straight and hit the impaction substrate, whereas smaller particles continue to move with the bending aerosol jet stream. The large particles are collected on the impaction substrate, and the small ones remain airborne. As the particles strike the surface of impaction substrate, the larger particles tend to bounce; to minimize the particle bounce, high-vacuum silicone grease is applied on the impaction plate. Thus, an impactor consists simply of a nozzle, either round or rectangular in shape, and an impaction plate (Fig. 3.1).

According to the impaction theory, the cut-point of each impaction stage can be calculated by using the dimensionless Stokes number (S_{tk}), which is defined as follows:

$$S_{tk} = \frac{\rho_p d_p^2 U C_c}{9 \eta W} \tag{3.1}$$

Stokes number is the main impaction parameter which governs the collection efficiency. The theoretical d_{50}

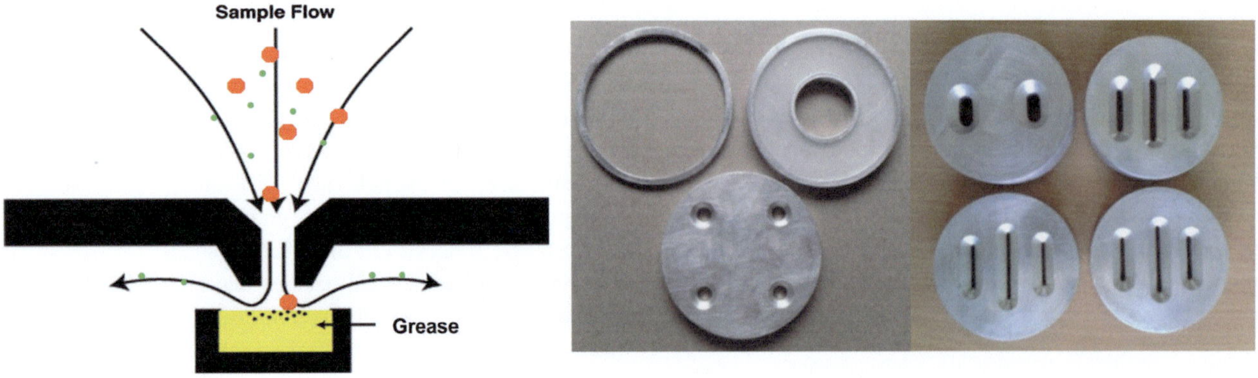

Fig. 3.1 Basic inertial impaction mechanism and various types of impactors [1]

(cut-point diameter) can be calculated by using the equation [2]:

$$d_{50}\sqrt{C_c} = \left[\frac{9\pi\eta D_j^3(S_{tk50})}{4\rho_p Q}\right]^{1/2} \tag{3.2}$$

This equation works fairly well for flat surfaces.

Studies have shown that an impactor can be characterized by the main impaction parameters which are dimensionless as *S*/*W* ratio (where *S* is the jet-to-plate distance and *W* is jet width or diameter) and jet Reynolds number (Re). Various studies have shown that similar shapes of experimental efficiency curves are obtained except for the case when jet Reynolds number is very low (Re < 500) and when it is extremely high (Re > 25,000). Also, the $\sqrt{S_{tk}}$ drops sharply with decreasing the jet-to-plate distance.

3.4 Materials and Method

3.4.1 Sampler Development

The impactor nozzles designed using the above-shown Eqs. (3.1) and (3.2) were tested in a laboratory setup as depicted in Figs. 3.2 and 3.3. Artificial aerosol was generated using sodium chloride (NaCl) solution for wet aerosol and dolomite for dry aerosol. A dry aerosol generator was employed using dolomite powder (sieved through a 45 µm size mesh) to produce a stable flow of polydisperse aerosol. A portable aerosol spectrometer (PAS 1.108, Grimm GmbH) was used to test the performance of the impactor.

Parametric investigation was carried out using impactor nozzles of different diameters, varying total flow rates, and space between impactor nozzle and substrate, in order to optimize the PM collection efficiency curve for the desired cut-point of 1.0 µm. High-vacuum grade silicone grease of depth 0.40 cm was used as an impaction substrate which allows particles to penetrate into it, and soon, their surface is wetted with silicone oil, hence eliminating particle bounce-off. About 2840 kg/m^3 of density (ρ_p) and 1.10 of shape factor (χ) of dolomite [1], respectively, were used to obtain aerodynamic diameter (d_a) for corresponding optical diameter (d_{op}) using the following relation:

$$d_a = d_{op}\left(\rho_p/\rho_o\chi\right)^{1/2} \tag{3.3}$$

The details of the dry aerosol generation system can be found in several earlier publications from our group [1, 3, 4].

3.4.2 Sampling Site and Procedure

This laboratory-evaluated sampler was next subjected to field measurement within the IIT Kanpur campus. The sampling was carried out for nearly 100 days spread evenly over a year-long period from 2008 to 2009. The sampler was placed on the roof of a building around 12 m high. The sampling period was 8 h from 9 am to 5 pm. Flow rate maintained through the sampler was 10 LPM using a needle valve and vacuum pump and monitored using a calibrated rotameter. The samples were collected on 47-mm-diameter PTFE (Teflon) filters (2 µm pore size, Whatman). Since the collected mass was of very low quantity, extreme care was taken during sampling and analysis of the samples to avoid any contamination. A microbalance (Mettler Toledo) with a LDL of 3 µg was employed for gravimetric analysis. The PTFE filters were chosen due to their chemical inertness, very little moisture absorption property, and overall stability to withstand changing weather conditions. All the filters were preconditioned at 25 °C and 60% relative humidity before sampling and post-conditioned after sampling at the same condition, and 10% of the total number of samples were always kept as a blank [5]. The blank filters were kept at the same condition as the actual samples. Later,

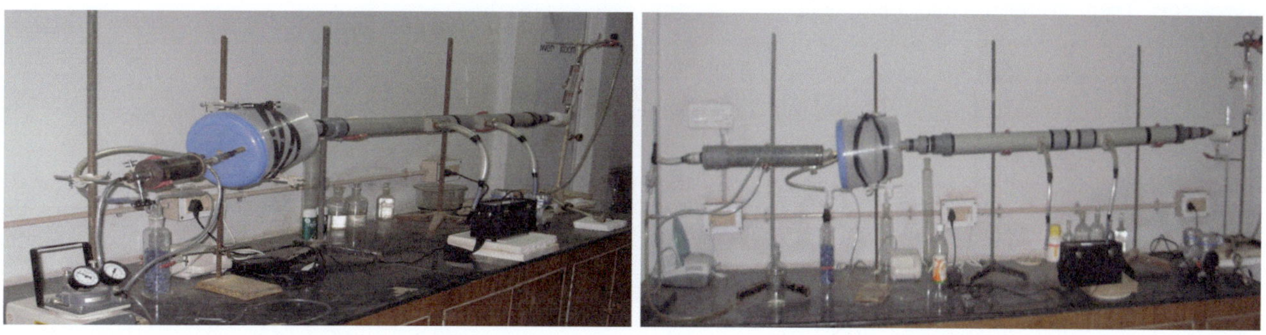

Fig. 3.2 Dry and wet aerosol generation system and impactor testing rig

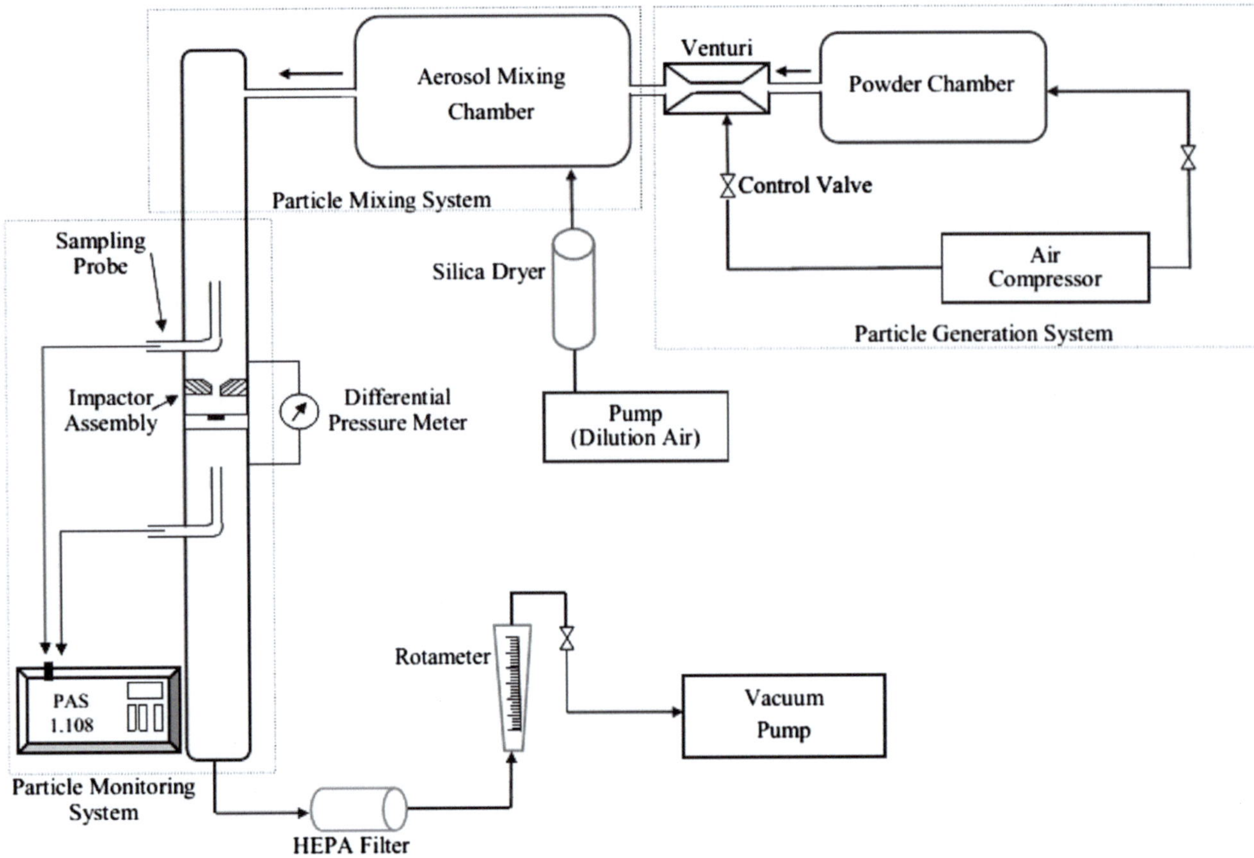

Fig. 3.3 Experimental setup for impactor characterization [1]

the sampling filters were immediately transferred to a sealed plastic container and kept in refrigerator till further analysis.

3.4.3 Chemical Analysis

Chemical analysis of the exposed filters was carried out in the following way: Each exposed filter was cut into two equal portions. The first half portion was cut into several small fragments using a plastic scissor and kept in a digestion vessel. Then, 20 ml of conc. nitric acid (70%, Supra-Pure, Merck GR grade) was poured into the digestion vessel and placed over a hot plate for around 2 h at 180 °C. The residual was then filtered through 0.22 µm Teflon filter and diluted using MilliQ water to 50 mL total volume and subjected to elemental analysis. Each blank filter was digested in the same way as the sample filters. The elemental analysis was performed using an Inductively Coupled Plasma Optical Emission Spectrometer (ICP-OES, iCAP-duo 6300, Thermo Inc.). The other half of the filter was subjected to aqueous extraction using ultrasonication and subsequent analysis by ion chromatography (Metrohm, Compact IC 761). PM collected on filter samples was analyzed for major ions: NH_4^+, F^-, Cl^-, NO_3^-, SO_4^{2-}, and PO_4^{3-} [3, 5].

3.5 Results and Discussions

3.5.1 Stability of the Aerosol Generation Systems

Figure 3.4 shows the size distribution of the wet and dry aerosol generated for calibrating and evaluating the PM_1 impactor. Both the systems generated stable size distribution for nearly 40 min, a sufficient time duration to carry out various parametric investigations for the impactor characterization.

3.5.2 Developed Sampler

Figure 3.5 depicts the internal components of the PM_1 sampler including rain cover, impactor nozzle plate, substrate plate, filter holder ring, and base part, and also the assembled complete sampler as well as its field setup. In addition, several other key accessories such as a needle control valve, tubing, calibrated rotameter, and a vacuum pump are required for proper use of this PM_1 sampler. This has been designed to have a 1-µm cut-point when operated at an air flow rate of 10 LPM [3].

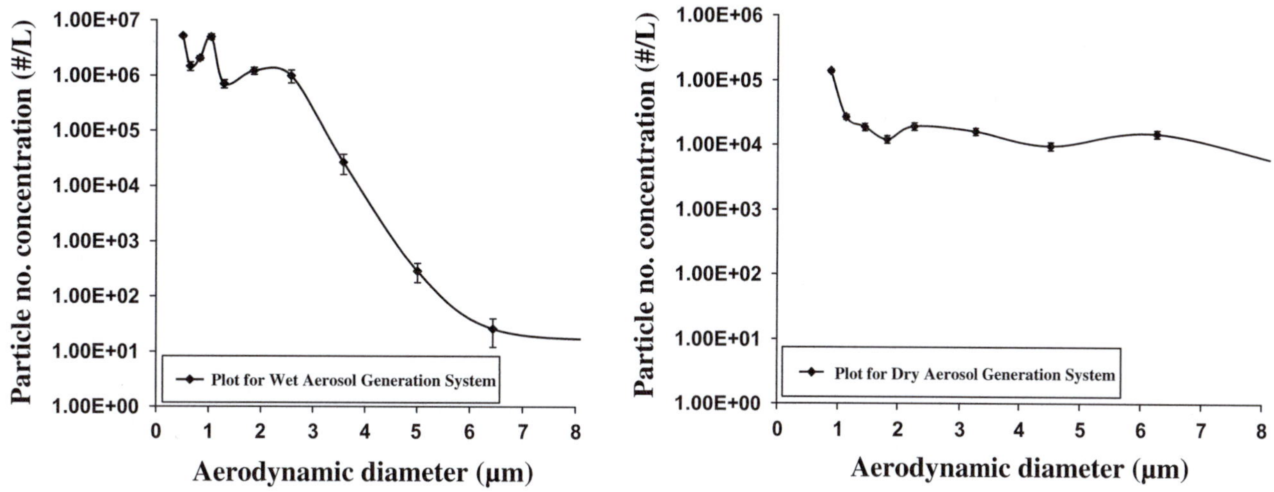

Fig. 3.4 Performance of the wet and dry aerosol generation system

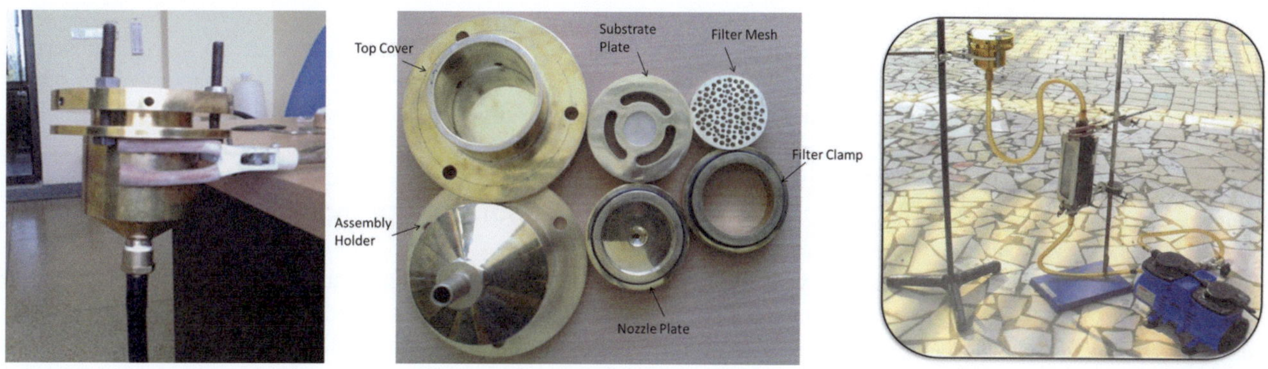

Fig. 3.5 Developed $PM_{1.0}$ sampler with its internal components and its field setup

The collection efficiency curves obtained for the optimized impactor nozzle using both wet and dry aerosol generator are shown in Fig. 3.6. There is a reasonably good agreement between the two curves. From the plot, it can be concluded that cut-point diameter (d_{50}) is at 1 µm [3].

Table 3.1 provides the details about the design parameters and performance of the optimized $PM_{1.0}$ impactor nozzle.

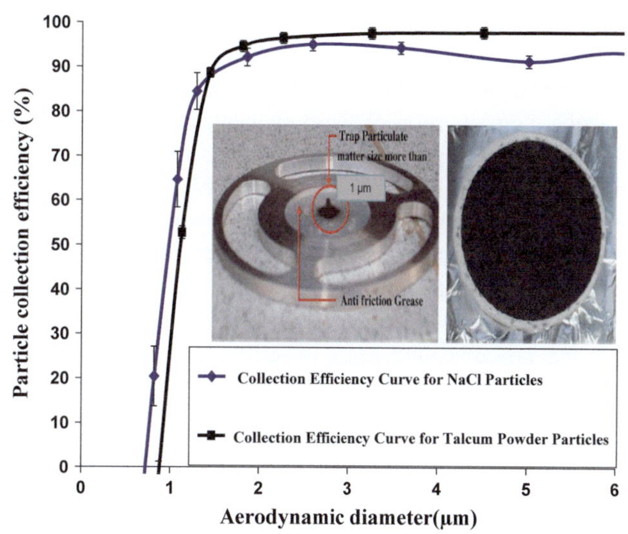

Fig. 3.6 Collection efficiency curves of $PM_{1.0}$ impactor [3, 4]

3.5.3 Sampler Field Evaluation and $PM_{1.0}$ Mass Concentration

The PM_1 concentration in the samples varies significantly from monsoon to summer season. During monsoon season due to high relative humidity, heavy rainfall, and less turbulence in the atmosphere, PM_1 concentration was found to be significantly lower as compared to the other seasons [5].

Table 3.1 Characteristic features of the submicron sampler [3]

Physical characteristics	
Nozzle width (*W*)	3.2 mm
Nozzle-to-plate distance (*S*)	2.5 mm
Reynolds number (Re)	4424
Experimental results	
d_{50} or cutoff diameter	1.0 μm
σg (GSD)	1.2
ΔP (pressure drop) using a 47-mm Teflon filter	18.5 cm of H_2O

Table 3.2 Seasonal average PM1 mass concentrations measured at IIT Kanpur [6]

Season	Mass concentration (μg/m^3)
Monsoon	30.1
Post-monsoon	63.8
Winter	199.0
Pre-summer	77.1
Summer	142.3

Elements such as Ca, Mg, and Fe that are mainly of crustal origin and mostly associated with coarse-medium size range were found to be higher in summer season possibly due to resuspension of crustal dust. Table 3.2 depicts the average PM_1 concentration measured for the different seasons at IIT Kanpur.

Overall, 100 days of samples were collected (including blanks) spread over the entire duration of a year. Elements such as Ca, Mg, and Fe that are mainly originated from crustal sources were found to be higher in summer season due to higher wind speed which causes the resuspension of crustal dust and lowest during monsoon season when high humidity and rainfall prevented the resuspension of crustal dust. Whereas during winter season when the atmosphere is most stable with almost no turbulence, concentrations of these elements were still higher indicating that in submicron fraction, some portion of these elements may also be contributed by anthropogenic sources such as vehicular exhaust. Elements such as As, Cd, Cr, Cu, Zn, Ni, and Se that are mainly originated from anthropogenic sources (vehicular exhaust, coal burning, industrial processes, etc.) were found to be higher during winter season and lowest during summer period. In winter, increased anthropogenic activities (field burning, space heating, etc.), lower mixing height, and low wind speed and temperature lead to poor dispersion conditions which increases the concentration of those anthropogenic elements whereas in summer, better dispersion conditions lead to lower concentration for those elements [3, 5, 6]. IIT Kanpur is, in general, devoid of any industrial activity. The overall vehicle density too is very low within the campus. Still, very high concentrations of measured submicron particles point toward its transport via wind, and gas-to-particle formation of secondary aerosol aided by poor dispersion of the pollutants especially during the winter time.

3.6 Laboratory to Industry

An indigenously designed and developed sampler to collect submicron aerosol was tested for its performance in the field. After completion of its initial performance evaluation in 2010, several publications related to this work appeared in leading international journals in the field of aerosol science and technology [3–5]. Besides, many scientists used this sampler for their research work. However, no patent application could be filed on this developed technology as we had already published this work. In early 2014, the owner of Envirotech Instruments Pvt. Ltd. visited IIT Kanpur and had a day long discussion on this sampler, and various studies carried out using it. They were very keen to bring this technology to the Indian market as it had the potential to be sold at nearly 15–20 times lower price than its available international counterparts. In January 2015, a formal MOU was signed between IIT Kanpur and Envirotech Instruments Pvt. Ltd. providing a 10-year exclusive technology license to the company in lieu of a token licensing fee. IIT Kanpur provided sampler design and key features to the company which then fabricated 5 such samplers. These samplers were tested by us, and a detail report was provided to the company. After several iterations and deliberations and nearly a semester long co-location testing of PM_1 sampler with other well accepted reference standards, the final version (APM577) was formally launched in the market on October

Fig. 3.7 Commercial version (APM 577) of the submicron sampler

1, 2015. The journey of the PM_1 sampler right from its inception, design, evaluation, usage, and then to its commercial launch in the marketplace has been a long but satisfying one for all those involved with it.

3.7 Commercial Sampler Highlights

The APM577 (Fig. 3.7) is a manual method for sampling very fine submicron particulates (PM_1 fraction). This sampler is based on impactor design standardized by Indian Institute of Technology, Kanpur. The air enters in the system through an omnidirectional inlet designed to provide a clean, aerodynamic cut-point for particles. The air sample and fine particulates exiting from the PM_1 impactor are passed through a 47-mm-diameter PTFE filter membrane that retains the submicron PM_1. The sampling rate of the system is held constant at 10 LPM by a suitable critical orifice. A dry gas meter is incorporated to provide a direct measure of the total air volume sampled. Time totalizer provides information on actual run time of sampling. An electronic timer is also provided to start the sampling and shut off the sampler as per the requirement of the sampling, thus eliminating the need of the presence of man power at site.

A provision is available to attach a Gaseous Sampling Attachment to collect sample for gaseous pollutants simultaneously. The system is provided with a brushless, continuously rated induction motor-driven suction pump which eliminates electromagnetic interference (EMI) in gadgets such as TVs, which sometimes hinders operation of air samplers in residential areas and is immune to voltage fluctuations. APM577 sampler is housed in a compact, light, robust, and weatherproof cabinet made of aluminum.

Acknowledgements The main funding source for the development of the PM_1 sampler was the faculty initiation grant provided to Dr. Tarun Gupta by IIT Kanpur. I would like to thank many of my previous and current students: Kamal, Abhishek, Anil, Anand, Daya, Shefali, Amit, and Dharmendra. Special thanks are due to the staff of SIDBI Innovation & Incubation Centre (SIIC) at IIT Kanpur.

References

1. Kumar A, Gupta T (2015) Development and laboratory performance evaluation of a variable configuration PM1/PM2.5 impaction-based sampler. Aerosol Air Qual Res 15:768–775. doi:10.4209/aaqr.2014.11.0307
2. Hinds W (1999) Aerosol technology: properties, behavior, and measurement of airborne particles, 2nd edn. Wiley, Interscience
3. Gupta T, Chakraborty A, Ujinwal KK (2010) Development and performance evaluation of an indigenously developed air sampler designed to collect submicron aerosol. Annals Indian Nat Acad Eng (INAE) 7:189–193
4. Gupta T, Jaiprakash, Dubey S (2011) Field performance evaluation of a newly developed PM2.5 sampler at IIT Kanpur. Sci Total Environ 409:3500–3507. doi:10.1016/j.scitotenv.2011.05.020
5. Chakraborty A & Gupta T (2010) Chemical characterization and source apportionment of submicron (PM1) aerosol in kanpur region, India. Aerosol Air Qual Res 10:433–445
6. Singh DK, Lakshay, Gupta T (2013) Field performance evaluation during fog-dominated wintertime of a newly developed denuder-equipped PM1 sampler. Environ Sci Pollut Res 21:4551–4564. doi:10.1007/s11356-013-2371-9

Author Biography

Dr. Tarun Gupta is currently serving as a professor in the Department of Civil Engineering, Indian Institute of Technology, Kanpur. He earned his doctoral degree in Environmental Health from Harvard University, School of Public Health, USA, in the year 2004. He received his M.Tech. in Environmental Science and Engineering from Indian Institute of Technology, Bombay, in the year 2000. He has authored more than 65 journal publications and filed applications for 2 Indian patents. His published work related to changes in critical lung function parameters to acute air pollution is a first of its kind for Indian population. He has developed several particle monitoring devices at IITK. He has been bestowed upon with salient recognitions such as the NASI SCOPUS Young Scientist Award (2015), P K Kelkar Research Fellowship (2015), INSA Medal for Young Scientist (2011), INAE Young Engineer Award (2009), and IEI Young Engineer Award (2008).

Manning the Unmanned—The Aarav Story

4

Vipul Singh

Abstract

Aarav Unmanned Systems is a start-up company designing and developing cutting edge technology in the field of unmanned aerial vehicles (traditionally known as drones). Aimed at establishing itself as an end-to-end solution provider offering its products, services and extensive analytical support, the company has already emerged as the first Indian company to offer indigenously built technology in the national market for civil applications. This article provides an overview of the company's vision, its major offerings and the future focusses.

Keywords

Unmanned aerial vehicles (UAVs) • Drones • Quadrotors • Nayan

4.1 Introduction

Land/topographic survey is one of the most primary activities undertaken to plan, execute and monitor all kinds of infrastructure and resources development. Traditional methods of land surveying which are widely adopted, especially in India, pose several challenges. Some of these include poor accessibility in densely populated areas and hilly terrains, slow and labour-intensive, expensive equipments, limited graphical output and prone to human errors and data inaccuracies. With the increasing demand for speed, efficiency, accuracy, safety and better project management, land surveyors are always seeking for faster, more flexible, more accurate, safer and enriched way of collecting the survey data (not to mention far better representation of this data).

UAV (Unmanned Aerial Vehicle) technology has an edge over land-based systems. UAVs are terrain independent, several times faster than the land-based systems, yield million times richer data and 3D interactive. When combined with photogrammetry, it generates accurate and detailed digital elevation models or 3D representations of the terrain surface, thus transforming and changing the way decisions are made in multiple sectors. Though the Indian market is still unaware of this emerging technology, it is slowly waking up to the inherent potential and capabilities of this technology. AUS (Aarav Unmanned Systems) realized the potential of the technology and entered the field with the determination of creating a whole ecosystem for organizations, which consists of its products (hardware and software) that capture highly sophisticated data and services that analyse this data.

4.2 Aarav Unmanned Systems (AUS)

AUS started its journey in 2013 with three of its cofounders coming together to represent IIT Kanpur in an SAE competition organized by NASA wherein they had to design a UAV that weighed least and lifted maximum payload. During this visit to the USA, they understood the true potential of UAV technology for Indian market. After a lot of brainstorming on possibilities versus fantasies, they realized that the utilization of UAVs for engineering applications is the future of this technology. That was the point when they started working on building technology for 3D

V. Singh (✉)
Aarav Unmanned Systems Pvt. Ltd, D-15/49 Dashashvmedh Ward-3, Varanasi, 221001, UP, India
e-mail: vipul@aus.co.in

B.V. Phani and S. Khandekar (eds.), *Innovation, Incubation and Entrepreneurship*, IITK Directions 1, DOI 10.1007/978-981-10-3334-6_4

mapping and GIS surveying. The same year the trio incubated their company at SIDBI Incubation and Innovation Centre (SIIC) at IIT Kanpur considering it to be the best place in terms of infrastructure and laboratories required to build technology like theirs. Also, the presence of a dedicated airstrip in the premises to test (and crash) their drones during the development phase came as a bonus.

Today, AUS has successfully ventured into civil, research and defence application capabilities using short-range UAVs. It is focusing on designing most advanced drones that are customizable based on the requirements of different sectors. It is also involved in the R&D of control systems, sensors and other hardware and software systems. The company intends to focus equally on the UX/UI besides building technology in order to churn out products that are efficient and very simple to use.

4.3 Technology

At AUS, UAV technology is combined with the best photogrammetric practices to generate accurate and detailed digital elevation models (DEMs) or 3D representations of the terrain surface. Currently, they are producing topographic data with a level accuracy of 50–60 mm which is comparable to traditional GPS ground-based techniques.

To create 3D maps from aerial photogrammetry, the camera is mounted on the drone and is usually pointed vertically towards the ground. Multiple overlapping photographs (80–90% overlap) of the ground are taken as the UAV flies along an autonomous programmed flight path called a waypoint. Overlapping photographs of an object or land by 80–90% would be impossible to complete accurately by pilot navigation. It is therefore essential to have a UAV which has an autonomous waypoint navigation technology.

In the case of theodolites, two angles are measured to generate a line from each theodolite. In the case of photogrammetry, it involves estimating the three-dimensional coordinates of points on an object/land employing measurements made in two or more photographic images taken from different positions. Common points are identified on each image. A line of sight (or ray) can be constructed from the camera location to the point on the object. It is the intersection of these rays (triangulation) that determines the three-dimensional location (X, Y, Z) of the point. More sophisticated algorithms can exploit other information about the scene that is known a priori, for example symmetries, in some cases allowing reconstructions of 3D coordinates from only one camera position.

Images captured in sequence & example calculation of image positions, ground measurements.

In order to achieve (and improve) these accuracies, AUS follows purely scientific methods that include extensive new age research and methodologies developed indigenously. A lot of on-board UAV sensor data (viz. attitude, location and state) are continuously stitched with images to obtain good results and accuracies.

4.4 Products and Services

The UAVs designed and developed by AUS can be deployed into several applications such as GIS surveying, 3D modelling, precision agriculture, institutional research, industrial inspection, thermal inspection, aerial photography, disaster/event management and Homeland Security and Defence Systems. However, the current focus is on using UAVs for topographic surveying and precision agriculture. AUS provides cost-effective and 30X faster UAV-based land surveying solutions for civil engineering industries and utility industries. These in-built UAVs are capable of:

- Topographic surveys,
- Volumetric surveys,
- Quarry/landfill surveys,
- Construction progress surveys,
- Site design,
- Flood risk surveys,
- Route surveys, and
- Environmental impact assessments.

4.4.1 Nayan 1

Nayan 1 is a high-performance quadrotor used as a research and development platform for vision-based navigation and control algorithms. The product can be customized with any appropriate payloads. Its main features include the following:

- Autonomous mission capable.
- Multiple fail-safe options.
- Two 90 fps global shutter CMOS camera for stereovision.
- GHz octacore on-board processor with 2 GB RAM.
- Dual processor autopilot to enable parallel processing of control system.
- Optic flow camera, stereo cameras and sonar to enable indoor position holding without GPS signal.
- Smart and rugged carbon fibre structure to enable easy use.
- Mote to enable multiple sensor integration and swarming.

Specifications	
All up weight	2200 gm
Max payload	800 gm
Range	LOS—1 km
Max altitude	3 km
Flight time	15–20 min

Nayan 1 Flying

4.4.2 Miscellaneous

Other technologies developed at AUS include the following:

- Multirotor autopilot with autonomous mission capable flight control system.
- Small fixed-wing autopilot with autonomous mission capable flight control system.
- Dual processor autopilot with one open computer for parallel processing of additional algorithm.
- FDAS-450, a flight data acquisition system for small-sized manned and unmanned aircraft.

Besides, the company is expecting to release a Hexa (with 6 arms and higher specs) and a fixed-wing UAV in next 2 months.

CASE STUDY—KUMBH MELA EXPERIMENT

Kumbh Mela is the largest religious festival and involves the pilgrimage of millions of Hindus to a sacred river. Unfortunately, many of the previous events have been marred by accidents and deaths. Human crowds are complex systems, and so predicting their behaviour is challenging. Therefore, a project has been undertaken to understand and manage huge human crowds. It aims to deliver the core components of an entire crowd management solution, all the way from designing and building personnel devices for tracking movement, to developing advanced computational models to help predict how the crowd may evolve. The project addresses three key research areas related to the understanding and management of huge human crowds: data collection, data analysis, modelling and prediction. The project is funded by the Netherlands Organisation for Scientific Research and Ministry of Communications and Information Technology, Government of India.

This year the festival was organized in Ujjain from April to May 2016. Aarav Unmanned Systems was chosen as one of the industrial partners for it and played a pivotal role in the project.

Mapping of Ujjain City—In February 2016, AUS mapped about 6 Sq km area of Ujjain city using self-owned UAV technology. The area of interest included roads, railway lines, ponds, populated areas, river and agricultural areas.

A part of Ujjain-Agra Road mapped as part of the whole project 5 GCP locations marked with white circles.

A total of 5 GCPs were established (as shown in the above image), and their coordinates were collected using a Trimble R10 GNSS system. As the project gets large, the number of overall GCPs required per unit area would exponentially come down, thus reducing the on-field execution time. The UAV system was fed with the KML file, and appropriate resolution settings were done. The flight plan and data capture scheme was automatically generated as shown in the following image:

Automatically generated flight plan and data capture schematic.

The flight (data capture) was completed within 12 min. The Captured data were retrieved from the SD card and fed into a third-party photogrammetry post-processing software. After pointing out the GCPs and feeding their coordinates in the software, it took about 2 h for the overall processing to complete. Following are the sample data generated for the corridor:

3D point cloud | The map consists of millions of points each having X,Y,Z coordinates.

Georeferenced Orthomosaic* | *<5 cm resolution with high clarity.

The provided model was of the resolution of 6 cm/pixel and average 6 cm accurate. This model is to be further used by the scientists to simulate the movement of crowd in the lanes, roads, alleys, etc., captured in the map. The input to the simulation would be the crowd movement data captured using other devices during the occurrence of the Kumbh Mela. The simulation will help understand the crowd behaviour during such large gatherings and identify the parameters which can warn about the possibility of impending stampede or chaos in an area well before time so that preventive measures can be taken.

***Digital Surface Model* (DSM).**

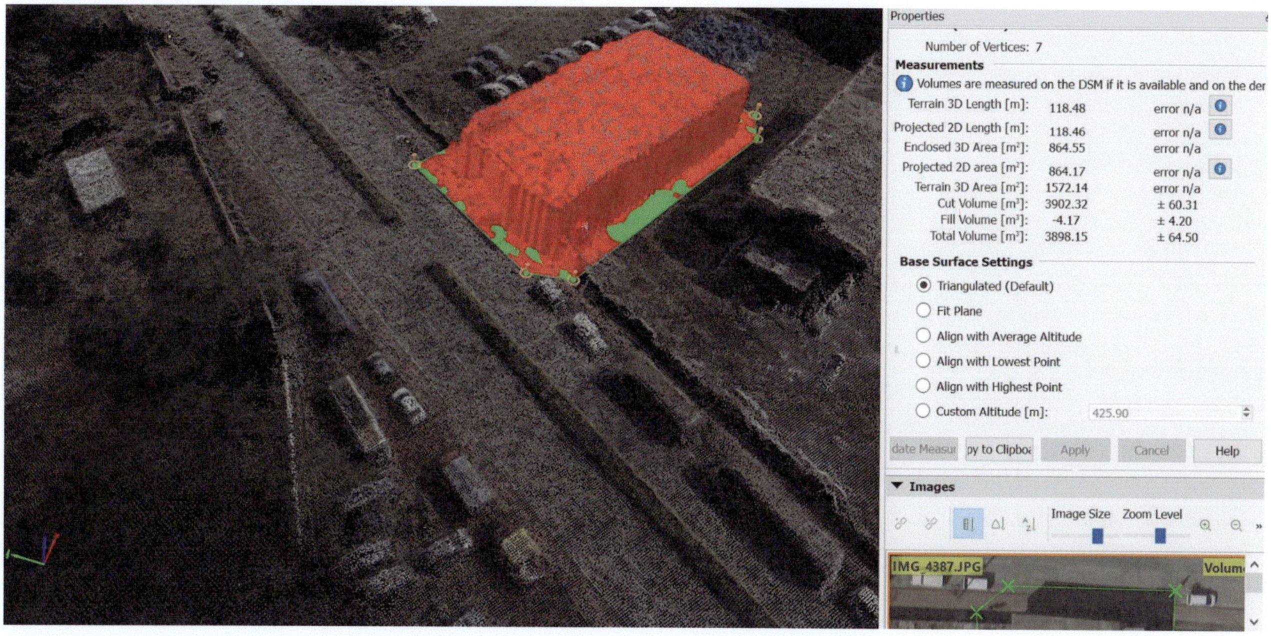

98% accurate Volume measurement within seconds.

4.5 The Road Ahead

AUS is slowly making its presence felt in the market. In a short span of time, it has already mapped more than 5000 ha as part of its mapping services and delivered its product Nayan to seven clients with several more offers being lined up. Despite several challenges such as tough regulatory impacts, import–export regulations and flying restrictions, the company has succeeded in raising an undisclosed amount of funding from StartupXseed, 3ONE4 capital (Mohandas Pai's family fund), The Phoenix Fund, SIIC and HNIs including Sanjay Jesrani and Ashok Atluri. The funds will be used for team and product development.

As a part of its long-term vision, the company wants to build more and more intelligent UAVs and venture out into unprecedented fields. Having already built its core technology base which took the company almost 2 years, it is now focussing on building expertise in machine learning and analytics to be in-line with its vision. It is highly confident that its advanced and affordable solutions will stand out in the international market as well where there is already a maturity amongst customers regarding this technology.

The 4 I's of Innovation

5

Sanjay Kumar, Pulak Bhushan and Shantanu Bhattacharya

Abstract

The 4i laboratory was set up in 2002 with the purpose of providing futuristic manufacturing solutions to the institute and the industry. Today, it intends to create a product-driven ecosystem that will boost the innovation and entrepreneurship culture within and outside the institute. This article provides a holistic overview of the laboratory in terms of its infrastructure and facilities and brings out its current role in supporting the manufacturing and designing work at the institute and the local industry. It also gives a plan document for the way ahead.

Keywords

Formula SAE race car • Dentist chair • Deskit • Autonomous navigation system • VLFM

5.1 Introduction

The 4i laboratory stands for the 4 I's of product design and development—Innovation, Integration, Incubation, and Implementation. The laboratory is envisaged as a central facility for concept design and product realization. It is involved in providing technical support and solutions to research students and sponsored research fellows thus helping them to convert their innovative ideas into working products or prototypes. Besides, it is heavily involved with undergraduate student technical clubs such as SAE and Robocon and helps in meeting out professional course requirements of students. The 4i laboratory has helped the institute in realizing many prestigious products such as Jugnu nanosatellite, hexapod, magnetic wall climber, Abhyast, low-cost dental chair, paper carry bag-making machine, and Deskit robotic arm, to name a few.

S. Kumar · S. Bhattacharya
Microsystems Fabrication Laboratory, Indian Institute of Technology Kanpur, Kanpur, India

P. Bhushan · S. Bhattacharya
Design Programme, Indian Institute of Technology Kanpur, Kanpur, India

S. Bhattacharya (✉)
4i Lab Coordinator, Indian Institute of Technology Kanpur, Kanpur, India
e-mail: bhattacs@iitk.ac.in

5.2 The Role of 4i Laboratory in Design and Fabrication

5.2.1 Aesthetic Design and Fabrication

Technological advances in the field of aesthetic designing have resulted in high-functionality, inexpensive, and easy-to-operate machines. A few years ago, laser cutting systems changed the face of engineering, aesthetics being one of the important aspects. Initially, machines were incapable of cutting complex designs with sharp angles in their geometry. But today's systems have successfully evaded these problems. It has been possible due to the controlled movement of the laser owing to sophisticated algorithms and advanced software engineering. Laser systems offer ergonomic and aesthetic advantages that increase efficiency and consistency. Laser beam machining is a non-traditional material removal process which uses thermal energy to remove material. In this non-contact mode of machining, a laser beam of certain wavelength strikes the surface causing localized heating thus resulting in melting and vaporization

B.V. Phani and S. Khandekar (eds.), *Innovation, Incubation and Entrepreneurship*, IITK Directions 1, DOI 10.1007/978-981-10-3334-6_5

of the material. Molten materials (in case of metals) and burn particles (in case of wood) are ejected by the assist gas pressure. As a result, cutting edges are clean and without burr or dust formation. Extremely fine contours can be achieved depending on the diameter of the laser beam. Various laser parameters such as power, worktable speed, and frequency control the depth of the cut making it easy to machine workpieces of variable thickness. High level of precision and positional accuracy is attained since the movement of the worktable is governed by a CNC controller. The laser machine available in the 4i laboratory (Fig. 5.1) is well suited for cutting, boring, and engraving, and is capable of machining various materials. Some samples processed on the machine are illustrated in Fig. 5.2a–d.

5.2.2 Precision Design and Development

Precision designing is of prime economic importance for many advanced manufacturing processes. Precision engineering as defined by Nakazawa is a set of systematic knowledge and principles for realizing high-precision machinery. It covers processing of materials sized in the range of micro-/nano- (e.g., microneedles, microchannels) to macro- (e.g., satellite rocket launcher) dimensions. This includes precision processing of materials, information processing systems, control systems, and automated manufacturing systems containing CAD and CAM systems. The main objective is to design and develop standard manufacturing processes for the fabrication of precise machine tools, measuring devices, and their control systems. The 4i laboratory is equipped with an EZEE drill and a CNC wire-cut EDM center for precision machining. The EZEE drill enables to drill small holes up to 250 microns in hard metals using brass and copper electrode tubes. The range of drilling can be up to 300 mm. Wire electrical discharge machining is a non-traditional material removal precise machine. The existing machine in the laboratory uses water as a dielectric fluid and is equipped with a precision stepper drive control. These machining centers are used for working on hard materials with high precision requirements of drilling and cutting. They are capable of generating intricate shapes and profiles with high precision.

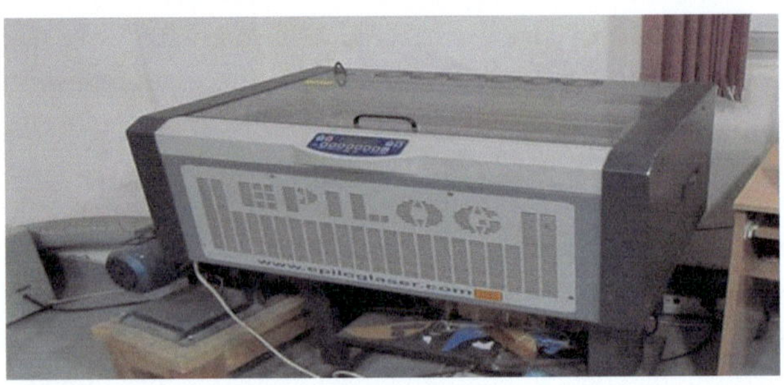

Fig. 5.1 CO_2 pulsed laser beam machine (Epilog Ltd)

5.2.3 Electronic Design and Fabrication

The motivation for fabrication of smaller circuits has been growing with the advent in manufacturing technology. The conventional ways of circuit fabrication are time-consuming, inflexible, and environmentally hazardous. Rapid prototyping enables design and manufacture of electronic circuits using rapid, better precision, high-performance, and eco-friendly techniques. Using this technology, lightweight and compactly integrated circuits have been fabricated for varied applications. Printed circuit boards (PCBs) are widely used as the basis for electronic circuits to provide mechanical support to electronic assemblies with circuit building capabilities. The PCB manufacturing process plays a central role in electronics and control systems. Several PCBs have been designed and fabricated for the institute and the incubative companies. The technology in the PCB manufacturing area is moving forward at a rapid pace and in order to keep abreast with the advancements, the 4i laboratory has, with support from SIIC, installed and commissioned a 3-D protolaser system where PCB etching can be carried out on an inclined surface. Among the other advancements in the area of PCB fabrication, the track sizes have substantially decreased and the numbers of layers in the boards have increased to accommodate the increased connectivity and high-density printing requirements. Single-layer PCB prototyping has been successfully performed for several projects. Various machines for the purpose of PCB laminating, prototype development, soldering, etc. are available at the 4i laboratory although there is an immediate need to reduce the track size and increase the accuracy of the PCB layout milling system. This can only be realized by increased institutional support and other in-house resource generation activities. The equipments used for PCB fabrication are shown in Fig. 5.3. The details of the PCB manufacturing machines available in 4i laboratory are provided below:

- Multilayer PCB laminating press.
- Circuit board plotters (LPKF Protomat 95S/II) for creating circuit board prototypes and engraving films as well as for engraving aluminum or plastic. This system needs augmentation that is planned in the future.
- Reflow ovens for lead-free Soldering (LPKF Proto-Flow S) for soldering, curing adhesives, and conductive polymers.
- Solder paste printer (LPKF ProtoPrint S) is an accurate manual stencil print for developing SMT prototypes and small batches.

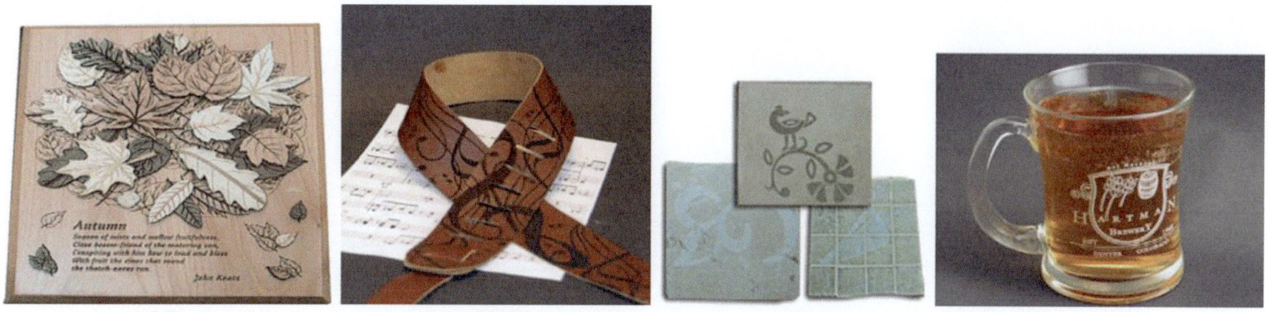

(a) (b) (c) (d)

Fig. 5.2 **a**, **b**, **c** and **d** CO_2 laser engraved samples on wood, leather, tile and glass, respectively, processed in the 4i laboratory

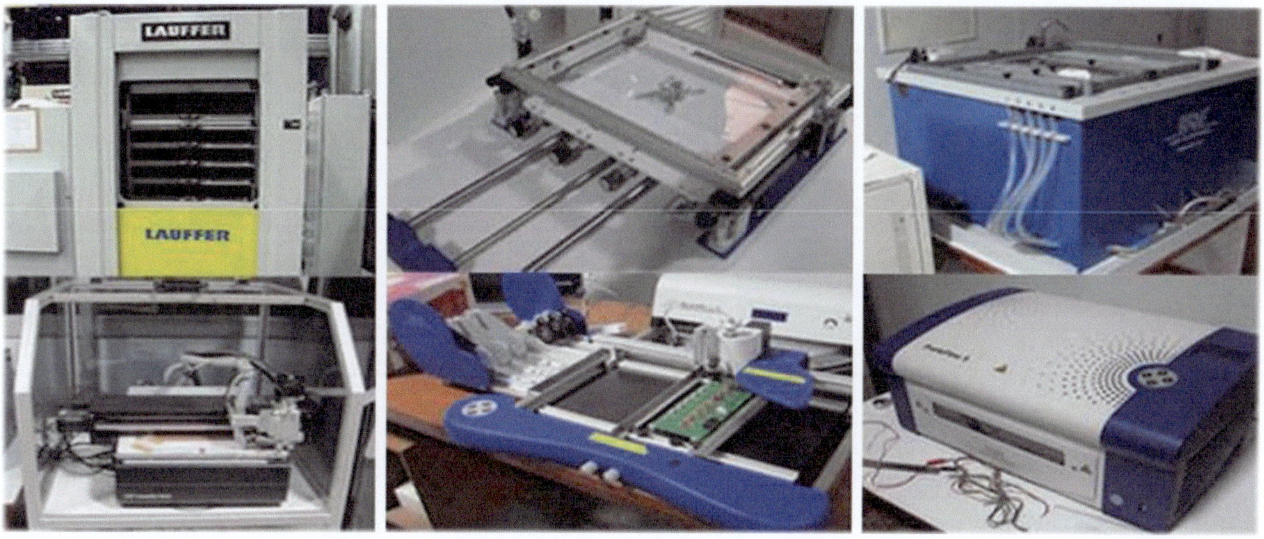

Fig. 5.3 Equipments used in PCB fabrication

- The through-hole plating systems (LPKF Contac RS) are used for electroplating of PCB prototypes in small batches. The system is equipped for chemical tin coating. A spray basin with external water supply assists in the PCB cleaning process.

Figure 5.4 shows some representative PCBs designed and developed within the 4i laboratory.

5.2.4 Reverse Engineering and Fabrication Using CAD/CAM Processes

Reverse engineering involves the transformation of real parts into engineering models and concepts. The developed models provide considerable information for the improvement in quality and ergonomics of the design, manufacture, and analysis. It recognizes the system's components and their interrelationships. Reverse engineering is commonly used in various fields such as automotive, consumer products, microchips, and mechanical designs. For example, design engineers first dissemble the existing product in the market and then technically innovate it in order to make the new developed product superior to the existing one. The 3-D laser scanner is one of the machines used for the reverse engineering purposes. It is a non-contact, nondestructive technology used to scan and create digital three-dimensional models of real objects by collecting data of their shape and appearance through a specialized UV imaging technique. It creates point clouds of data captured from the surface of the object that is being scanned. Common applications of this technology include industrial design, orthodontics and prosthetics, reverse engineering, and prototyping. Figure 5.5a shows a snapshot of the system, and the display shows the acquisition of scan data of a miniaturized human head which has been developed through rapid prototyping.

With a combination of the scanner data and the Delcam-driven auto-coding facility, a complete engineering

Fig. 5.4 **a** PCB fabricated for Technosum Pvt. Ltd (reprinted with kind permission from Technosum Pvt. Ltd). **b** PCB fabricated for Aarav Unmanned Systems Pvt. Ltd (reprinted with kind permission from Aarav Unmanned Systems Pvt. Ltd). **c** PCB fabricated for Technosum Pvt. Ltd. **d** PCB fabricated for Quad rotor(reprinted with kind permission from project Coordinator)

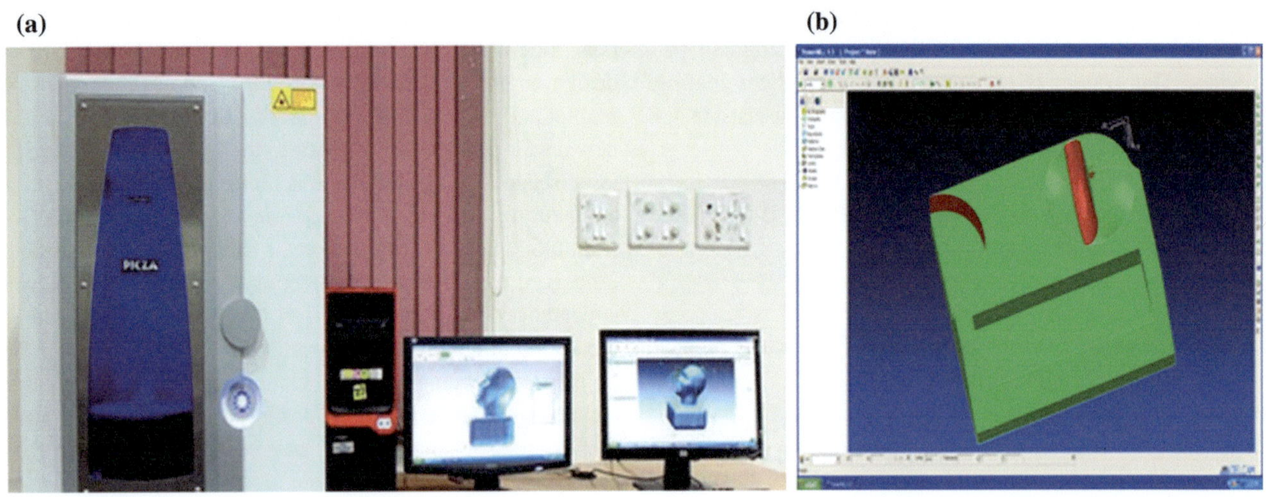

Fig. 5.5 **a** PICZA 3-D Laser scanner LPX-600 acquiring some images. **b** Delcam module performing auto-coding on scanned data of an engineering part

part has been reverse engineered for the demonstration purposes in training programs associated with institutional and outreach activities of the 4i laboratory (Fig. 5.5b).

Besides, the 4i laboratory has a variety of CNC turning and milling centers to fabricate the modified objects. It also houses a recently updated 5-axis system along with computer-aided manufacturing capabilities. Delcam and Mastercam softwares are used to generate the CNC code for complex profiles. They also help validate and plan the path movements before transferring them to the machine controller for actual processing. While Fig. 5.6a shows a photograph of the CNC turning center, Fig. 5.6b represents images of some representative parts manufactured by this machine.

The 3-axis milling machining within the 4i laboratory requires a simultaneous controlled movement of all the 3 axes, which is a basic requirement of most free-form surfaces. Complex operations, like the machining of sculptured surfaces required in molds and airfoils, fall into this category. Figure 5.7a shows a picture of this machine, and Fig. 5.7b shows the complexity of profiles that can be generated by this machine.

5.2.5 Free-form Design and Fabrication

Free-form surface geometry is controlled by a set of control points with different basic functions. There is no intrinsic notion for the free-form surfaces. The free-form surfaces contain one or more non-planar, non-quadratic surfaces. These surfaces fulfill the requirement of esthetic design. The applications of these geometries have grown rapidly in the field of aerospace, automobile, defense and consumer product industries. There are various methods or processes by which free-form surfaces can be generated such as CNC machining, rapid prototyping (FDM), and non-traditional machining (e.g., abrasive water jet machining).

Fused deposition modeling (FDM) is an additive manufacturing technology following a bottom-to-top approach for developing prototypes. It builds the final product by adding molten thermoplastic materials layer-by-layer. In a FDM process, CAD geometry is split up into layers and each layer is laid out in a sequential manner over one another to build the whole-part geometry. The working steps of the FDM process are summarized in Fig. 5.8a. In this machine, a thermoplastic wire is fed through heating filaments where it melts down in semisolid stage and is extruded out of a moving nozzle. The movement of the nozzle is controlled by programming software based on the layer-by-layer data generated out of CAD geometry. The extruded material gets deposited on the substrate and is immediately cooled and hardened. This process is comparatively a very slow process and hence is generally used for fabricating prototypes only.

The free-form contours can also be cut through abrasive jet machining process. In abrasive water jet machining process, a high-speed water jet with abrasive particles accelerates toward the workpiece and mechanically erodes a portion from the target area. The water jet movement is controlled by the CNC controller. The basic principle of material removal is through the mechanical fracture of the materials. The water jet washes away the fractured piece from the zone of machining. This process is widely used for brittle and hard materials such as glass, ceramics, concrete, and tough composites. Garnets, aluminum oxide, and silica (sand) are the most commonly used abrasive particles in these systems. The machine has the potential to cut composite materials approximately 10 times faster than the conventional machines. Figure 5.9a shows a schematic of

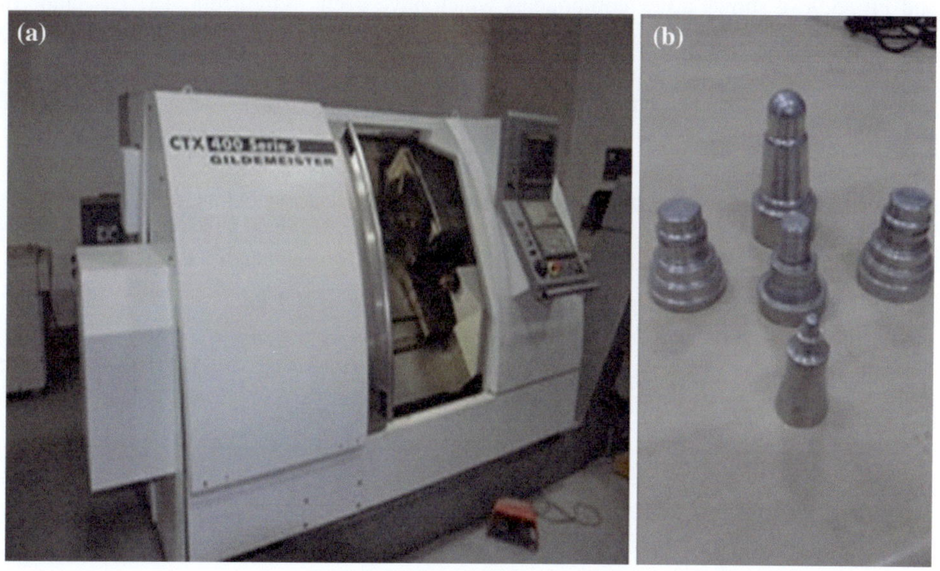

Fig. 5.6 **a** CTX Gildmeister Turning Center. **b** Representative samples made at 4i laboratory

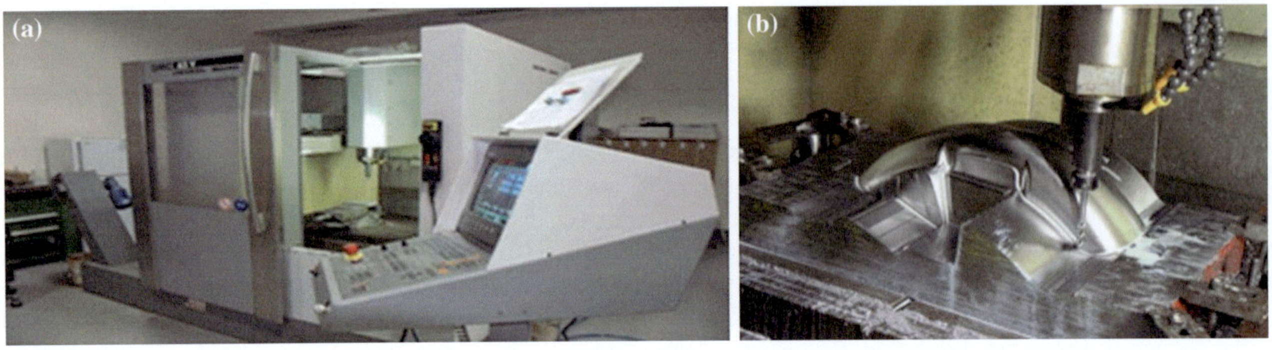

Fig. 5.7 **a** 3-axis CNC Milling Center (DMC 63 V Deckel MAHO). **b** Milling of a complex die shape being done at the 4i laboratory

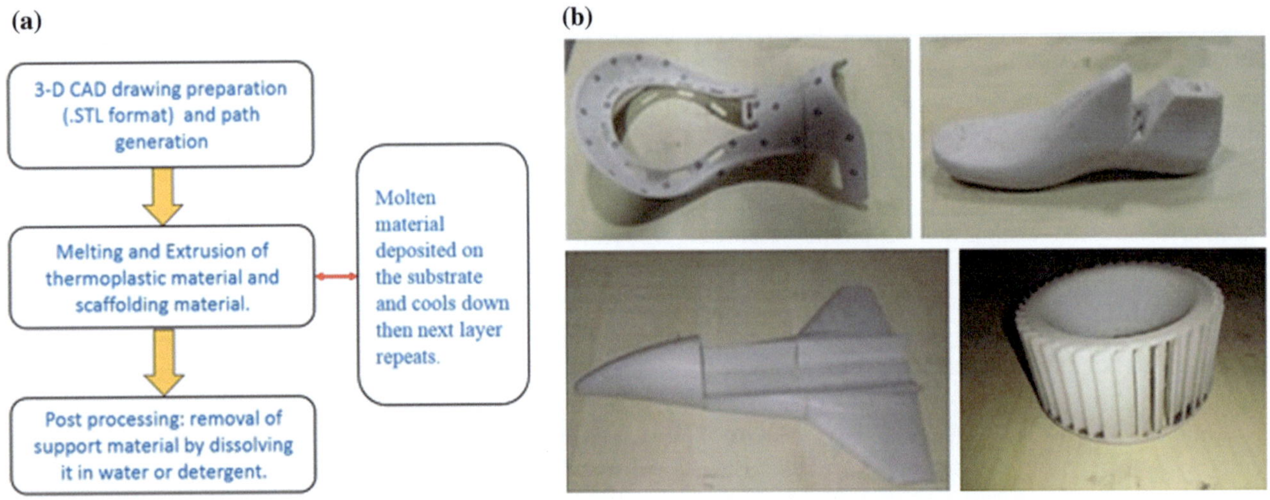

Fig. 5.8 **a** Working steps of fused deposition modeling. **b** Some representative samples made in the 4i laboratory

water abrasive slurry delivering system. The pressurized water is mixed with abrasive particles and flown through a nozzle kept at a certain standoff distance from the workpiece. Figure 5.9b shows a picture of the abrasive water jet setup, and Fig. 5.9c–e shows a few components which have been processed in the 4i laboratory on the water jet cutting system. Research efforts are also on to produce micropillars in metals using the water jet cutting system (Fig. 5.9f–g).

5.3 Foregoing and Completed Innovative Projects

In the 4i laboratory, several projects including industrial as well as institutional have been successfully completed and a few are still ongoing. The following section will illustrate some of the fabrication work done in the laboratory.

5.3.1 Formula SAE Race Car

Under the supervision of Dr. A.K. Agrawal, Dr. Tarun Gupta and Dr. Abhijit Kushari, the society of automotive engineers (SAE) club of IIT Kanpur has developed Formula-style race car. In 2013, IIT Kanpur's indigenously participated in Formula Student Italy. Some very critical parts of this Formula race car were fabricated at the 4i laboratory (Fig. 5.10).

5.3.2 Abhyast: Autonomous Navigation System (ANS)

Abhyast, Autonomous Navigation System (ANS), is a project funded by Boeing, USA. Abhyast is an autonomous navigation vehicle being undertaken as a joint activity

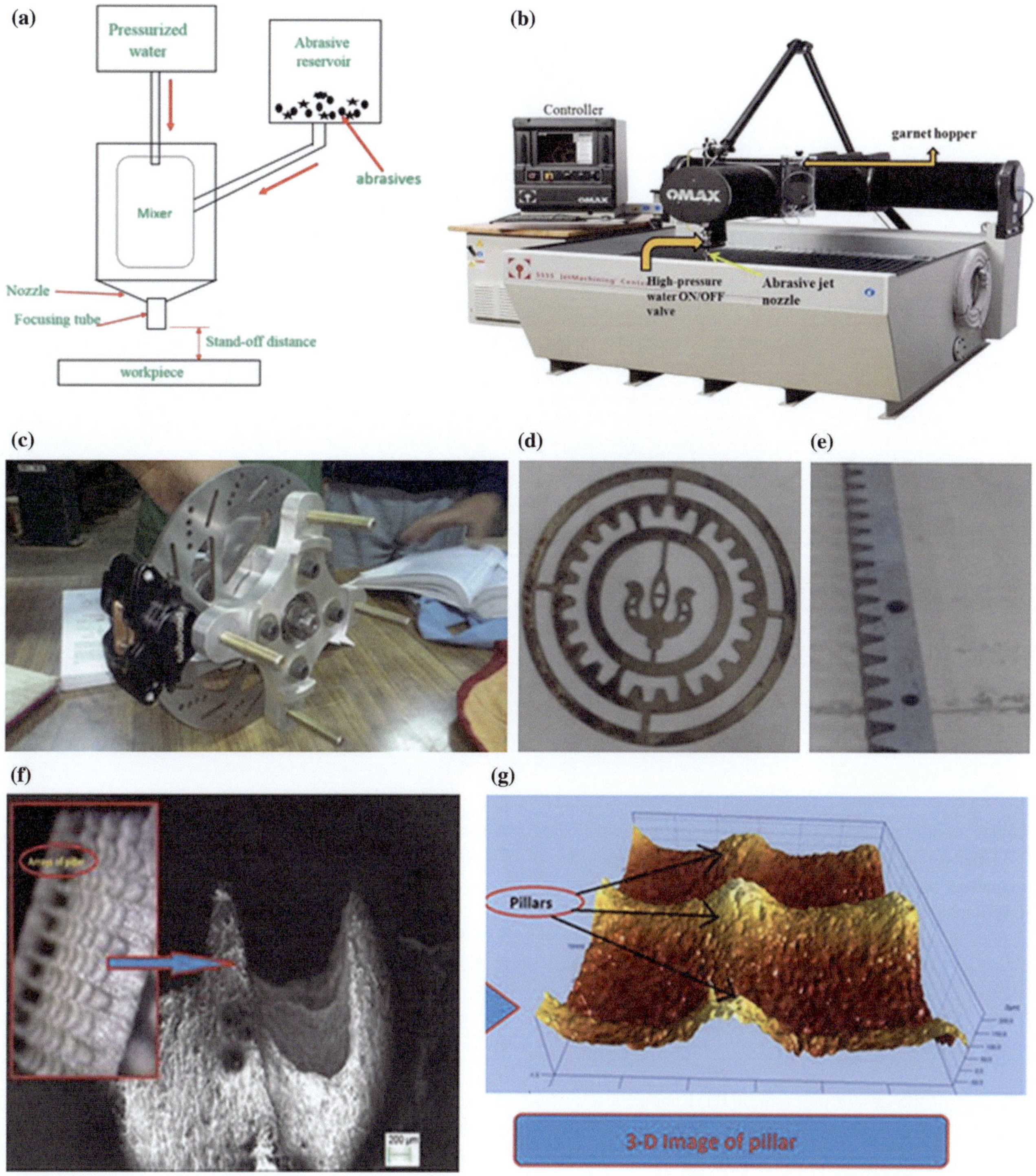

Fig. 5.9 **a** Schematic diagram of abrasive water jet delivery system. **b** Picture of the abrasive water jet system (OMAX 2652). **c** Brake assembly for SAE (on a workdone under coordinatorship of Prof. A.K. Agrawal, ME). **d** IITK logo. **e** Rack. **f** SEM image of micropillar arrays in metal and **g** Reconstructed image of the micropillars (on a workdone by Prof. S.K. Choudhury, ME and his research team)

between Indian Institute of Technology Kanpur and Boeing Corporation through their undergraduate relations program. The objective of this project has been to develop a robot which acts as an aid to the operator in carrying out specific tasks rather than an entity which is in need of continuous attention of the operator. It is a 30 cm * 30 cm * 30 cm robot capable of navigating from a specific source to destination using various local and global sensors. It is also

Fig. 5.10 Parts fabricated at 4i laboratory for the formula race car (reprinted with kind permission from project Coordinator, SAE)

efficient in communicating to the user using pre-established GSM networks. It also includes a semi-autonomous aerial navigation system which controls the ground robots for surveillance activities. Figure 5.11a–e shows a series of mechanisms which have been developed under this program.

5.3.3 Design and Development of a Low-Cost Dental Chair

As a part of the design program at IIT Kanpur, a few students along with a team of students from PUJ Colombia have designed a low-cost dental chair. The chair developed under the ME310 course program run at Stanford University has been named as 'Flux.' The prototype of the chair is currently commissioned at Sardar Patel Institute of Post Graduate Studies in Dental & Medical Sciences, Lucknow, where user trials are on. This has been hailed as a step toward providing low-cost dental treatment to masses. Figure 5.12a, b shows the chair that has been setup.

5.3.4 Deskit and Newspaper Carry Bag Machine

Recently, a group of students from the design program have developed an innovative solution to tackle the problem of low-infrastructure schools in rural India. The schools in rural India severely lack proper seating for the students. The product developed named DESKIT is a low-cost adjustable height school bag (Fig. 5.13a). The second product designed by the students is a low-cost paper folding machine capable of folding, creasing, and pasting newsprint to make paper envelopes. It has been realized keeping in mind the recent initiatives of the government to ban plastic carry bags that are environmentally unfriendly (Fig. 5.13b). Both these products have been developed with the fabrication support

Fig. 5.11 **a** Phase 1 and 2: rough terrain vehicle with obstacle avoidance. **b** CAD model of Phase 3 (ANS). **c** Phase 3: ANS with jumping capability. **d** CAD model development for ground vehicle. **e** Phase 4: wirelessly controlled Robot for disposing suspicious objects (under the supervision of Project coordinator, Dr. Shantanu Bhattacharya)

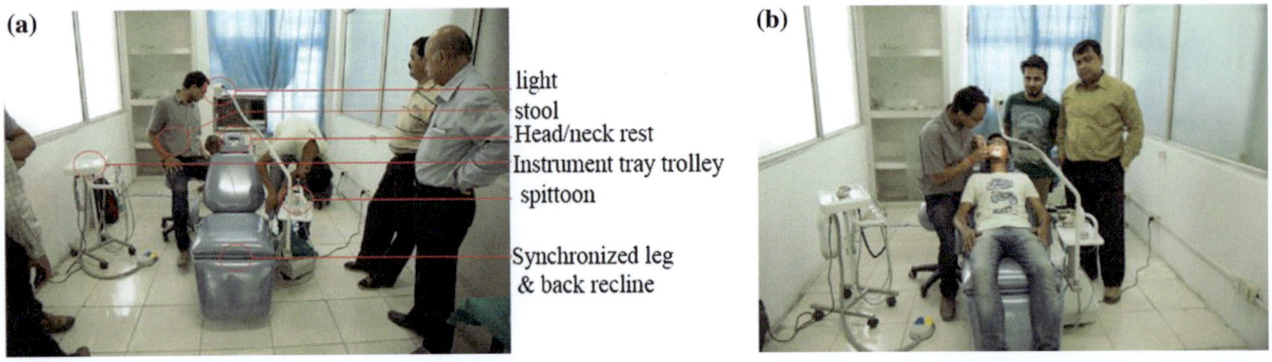

Fig. 5.12 **a** and **b** design of dental chair: dental chair is designed as a part of ME310 course at D School, Stanford University in collaboration with Sardar Patel Institute of Dental Sciences, Lucknow (reprinted with kind permission from project coordinator, Dr. Shantanu Bhattacharya and Dr. Himanshub Gupta)

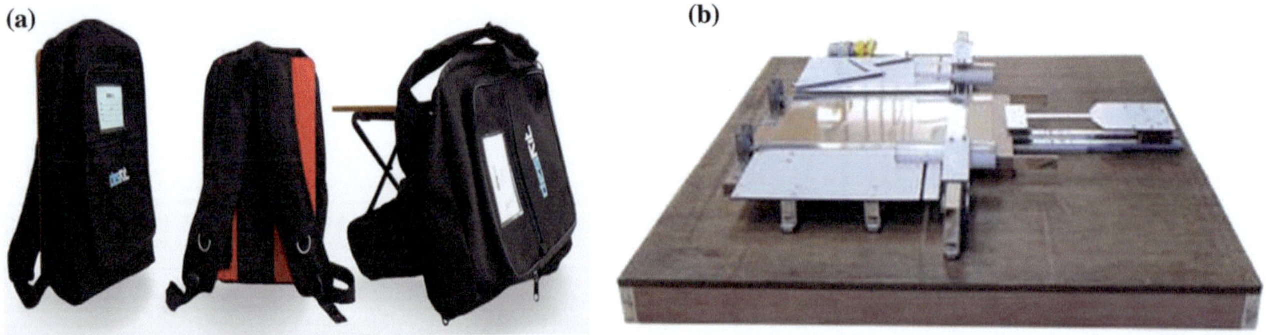

Fig. 5.13 **a** Deskit prototype. **b** Paper carry bag-folding machine prototype (reprinted with kind permission from Mr. Eshan Sadasivan, CEO, PROSOC Innovators)

from the 4i laboratory. The inventors of these products are launching their own enterprise named Prosoc Innovators.

5.3.5 UV-based Water Filter System

Another representative design project carried out through the fabrication support from the 4i laboratory is the solar UV water filter system. The filter removes particulate matter from water through a ceramic candle filter followed by a UV disinfection of the water sample. The filter is realized in two chambers, and an electronic PCB used for controlling the UV timer senses the water level and provides disinfection. Efforts are currently on to make this system solar operable (Fig. 5.14).

5.3.6 Unmanned Aerial Vehicle (UAV)

A low-altitude, long endurance fixed-wing unmanned aerial vehicle has been developed by a group of researchers at IIT Kanpur. The UAV has applications in surveillance activities related to the army, traffic control, and management and crowd monitoring activities. Parts from the main body of UAV have been fabricated at the laboratory using water jet cutting center, 3-D printing center, and CO_2 laser machining (Fig. 5.15).

5.4 Fabrication Support to the Institute

The 4i laboratory has been actively involved with several programs of the institute such as the Visionary Leadership Program in Manufacturing (VLFM) where the product design, product management, and computer-integrated manufacturing system modules seek the 4i laboratory's assistance. Figure 5.16a, b represents the paper bikes designed and fabricated in the 4i laboratory by VLFM students as a part of a paper biking design challenge. These bikes were made on paper with only 500 grams of some other material used and could handle one-person payload while being propelled by another person. The 4i laboratory has also provided support in the design and development of some modern foldable tandoors for the Indian kitchen (Fig. 5.16c, d).

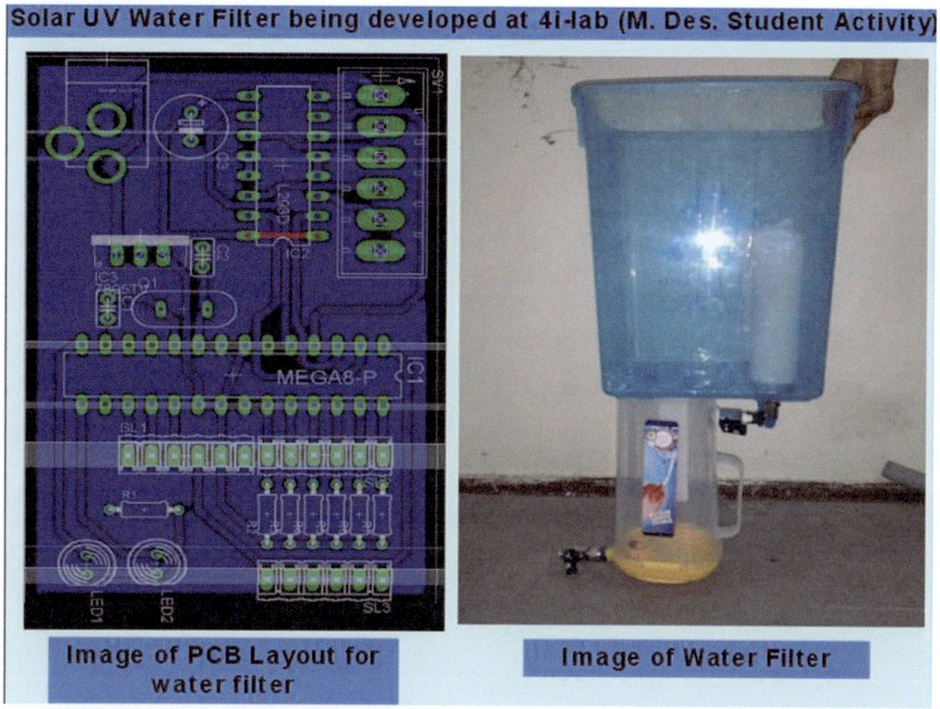

Fig. 5.14 UV water filter developed at the 4i laboratory (reprinted with kind permission from project leader, Dr. Shantanu Bhattacharya)

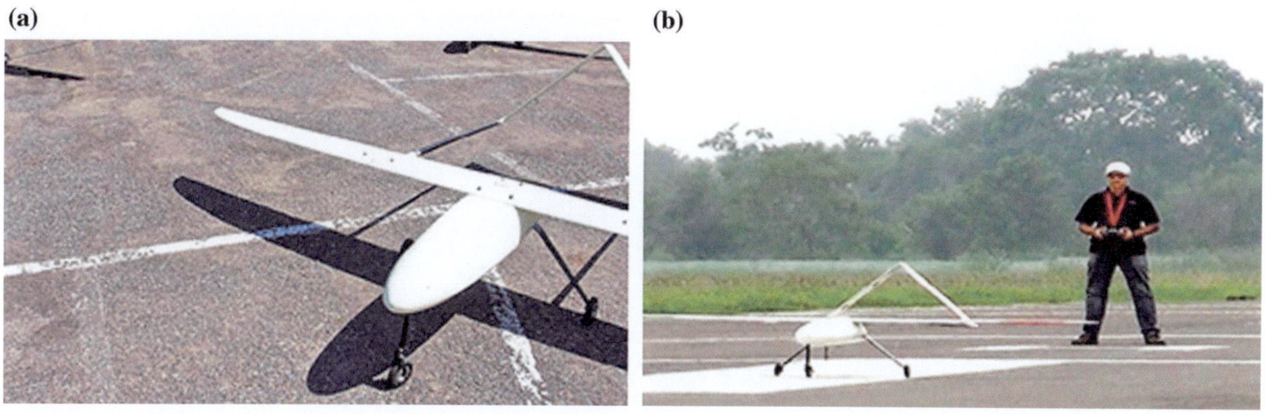

Fig. 5.15 **a** Developed UAV. **b** Test run of the UAV (reprinted with kind permission from Project Coordinators Dr. A.K. Ghosh and Dr. D. Philip)

5.5 Future Directions

The laboratory has been carrying out its vision through the varied activities mentioned above and hopes to intensify the existing activities. In line with this vision, a space called the student activity center has been formulated within the 4i laboratory where students from various clubs such as SAE and Robocon are provided seating and working space. A new area is being developed within the ambit of the 4i laboratory which supports the project activity from the Ministry of Textiles (MOT) in the area of brass castings. The laboratory, with financial support from MOT, will offer advanced training to brass manufacturing units through aiding brass clusters in the state. However, it needs some further infrastructural enhancements in terms of augmenting the accuracy of the PCB fabricating center and also getting another high-resolution rapid prototyping center where institutional support has already been requested. The 4i laboratory is committed to providing all kinds of advanced fabrication support to create a product-driven ecosystem within the institute.

Fig. 5.16 **a** and **b** some designed and fabricated paper bikes from the VLFM program with help and support from 4i laboratory **c** and **d** lowcost tandoors for Indian kitchen (reprinted with kind permission from VLFM course coordinators Prof. Jayanta Chatterjee and Prof. Shantanu Bhattacharya)

Acknowledgements The authors would like to express their gratitude to the students and faculty members who have been involved in performing fabrication activities at the 4i laboratory. A special acknowledgement goes to all the previous 4i lab coordinators who have supervised this lab. We would further like to deeply appreciate the financial support provided by various programs such as VLFM and MOT for carrying out projects described above. Needless to say a special thanks to the Dean of Research and Development for overall support for some activities. The authors would also like to especially thank the members of 4i laboratory Mr. Virendra Singh (4i lab in-charge), Mr. Santosh Paramanick, Mr. Ramendra Pathak, Mr. Vijay Pandey, Mr. Parmesh Kumar, Mr. Jashpal Singh etc. who have immensely contributed to the realization of the various modules as described in the above article.

The Game Changers—Seekers to Creators

6

Snehil Patel, Devin Haria and Kanupriya Agarwal

Abstract

A crucial step towards building a vibrant innovation ecosystem at IIT Kanpur was to energize and enthuse the student community with entrepreneurial zeal and to empower them to think out of the box and create, innovate and deliver. It all started with Magabucks event in 2000 and gradually took the shape of Entrepreneurship Cell (E-Cell) that today boasts of having supported 30+ start-ups from within and outside the institute. This article outlines the growth of E-Cell and the success stories that emerged out of this initiative.

Keywords

Entrepreneurship Cell • E-Summit • Techkriti • Start-ups

6.1 Introduction

Every year scores of young, exceptionally brilliant minds bustling with energy and riding high on confidence, join the institute. Most of the times, however, unaware of their innovation skills, they come with conventional mindsets harbouring age-old aspirations. The Entrepreneurship Cell (E-Cell) at IIT Kanpur is an effort towards instigating the entrepreneurial streak hidden in each of them and empowering them with the right set of tools to set up successful businesses. As a student initiative, it intends to build an enterprising community that is constantly adding innovative ventures to the Indian start-up ecosystem, *a community that sees modern day entrepreneurs as their idols*!

Like most other institutions where dynamism thrives, E-Cell has more faces to it than what just the name reveals. It believes that 'entrepreneurship' is not just about starting up and thriving a business. It is as much about taking risks, snapping at opportunities and evolving through mistakes as it is about making money. Aimed at bringing cultural and sociological transformation of the mindsets of students in a non-intrusive manner, E-Cell strives to preach that '***youngsters build their lives like they build start-ups***'. Borrowing from the title of Reid Hoffman's popular book, E-Cell wants 'you' to make a 'Start-up of You'. The evolution of this holistic and all-encompassing motto of E-Cell at IIT Kanpur has much to do with its journey since it came into existence.

6.2 The Birth

Conceptualized in the year 2000 by the pioneer batch of IIT Kanpur, Megabucks was IIT Kanpur's first formal tryst with entrepreneurship. The three-day international business and entrepreneurial festival was aimed at nurturing and inspiring business leaders of tomorrow. It served as a platform to bring together the start-up creators, the start-up promoters and the start-up supporters. Besides hosting a premier gathering of entrepreneurs, venture capitalists, trade associations, technocrats, incubation hubs, start-ups and the academic fraternity, it organized a wide range of business competitions that ruthlessly tested the young minds for their ability to ideate, prototype, conceive a business plan, market their ideas and woo investors. In a nutshell, three days turned out to be a training-cum-combat zone exposing students to the various facets of setting up and running businesses successfully. In 2008, there came a significant twist when Megabucks was dissolved and transformed into a 'technical and entrepreneurial' festival called Techkriti. However,

S. Patel (✉) · D. Haria · K. Agarwal
Indian Institute of Technology Kanpur, Kanpur, India
e-mail: snehilp@iitk.ac.in

B.V. Phani and S. Khandekar (eds.), *Innovation, Incubation and Entrepreneurship*,
IITK Directions 1, DOI 10.1007/978-981-10-3334-6_6

Megabucks had played its role by then. Entrepreneurship had become the 'buzzword' with over 12 companies being formed and officially registered during that span. A few of these such as Transversal e Networks, Ixigo and InMobi have emerged as big players in their fields.

Megabucks had set the stage for the next phase. The three-day event held once in a year could no more do justice with the budding enterprising culture and the growing demand for the year-round activities. IIT Kanpur needed the bearers of 'entrepreneurship', its supporters and promoters and voilà! This led to the birth of the Entrepreneurship Cell in 2009. The E-Cell was an entrepreneur's solution to promoting entrepreneurship. It was, rather is, as much a 'start-up' with its own set of vision and objectives, action plan and challenges, risks and opportunities.

6.3 The Growth

E-Cell was fed and nourished with new ideas and initiatives. A comprehensive structure was put in place with distributed roles such as marketing, corporate relations, media and publicity, finance and logistics to form a strong team. Its first few initiatives were its major ones that played a crucial role in setting up the arena for entrepreneurship within the institute premises. The events described below, besides scores of others, fed the fire of innovation that had been kindled by the birth of E-Cell. From the very beginning, there was an attempt to ensure that E-Cell had something for everyone, beginners, amateurs as well as seasoned professionals.

6.3.1 E-Summit

The annual flagship event, that exclusively caters to entrepreneurship enthusiasts from all over the country, has seen its 4th edition since its inception. With the theme 'Ready-Steady-Startup!', E-Summit's motto in 2015 was to reach out to the start-ups of tier-two cities of India such as Kanpur, Lucknow and Indore, which usually do not get the exposure like their counterparts in metropolitan cities. The whole event was meticulously designed to be a potpourri where different ingredients served and satiated different entrepreneurial needs.

UpStart, where start-ups go through multiple rounds of screening, which are essentially brainstorming sessions at different levels. Finalists get to develop their ideas and hone their business plans with a panel of mentors.

PuP (Pitch Ur Product) provides participants an opportunity to present their innovative ideas, including summer projects, as potential working start-ups in future and thus examine their ability to evaluate the market, customer needs and distribution strategies.

SoCha (Social Challenge) aim at directing entrepreneurial minds towards the social problems that plague the Indian society today and encourage them to provide a viable and sustainable solution for it.

Decrypt, the case study competition wherein participants are given a start-up case/crisis and have to step into the shoes of a founding member to present and defend their solution.

Rapidea, a fun event designed to test the problem-solving skills of an individual in a limited time frame—an essential attribute for any aspiring entrepreneur.

Networking session, a platform for start-ups to showcase their ideas among hundreds of potential users and investors with a realistic possibility of product acquisition. Networking arena is presided over by the head honchos of the industry and provides attendees with ample opportunity to network.

Crowded, styled as a pitch-cum-exhibition where shortlisted teams get 90 seconds to pitch their idea to the crowd, with an aim to get the crowd interested enough to visit their demonstration later.

Besides business competitions, a series of motivating talks by highly accomplished people are organised, which raise the entrepreneurial fervour within the institute to a new level. Aligned with the overall vision, E-Summit promotes the entrepreneurial spirit at all levels and convincingly propagates that entrepreneurship is for all, more like a lifestyle.

6.3.2 Start-up Internship Programme

The best way to popularize 'the start-up way' is to let students personally experience it. This thought instigated the E-Cell team to come up with the unique idea of running a 'Start-up Internship Programme'. SIP is exactly what its name suggests—start-ups hire students as interns. In its first edition in October 2010, SIP had only nine start-ups on board. Today, it registers over a 100 start-ups every year which hire more than 75 interns. Through SIP, E-Cell provides students an unprecedented opportunity to experience entrepreneurship first-hand and to connect with the 'to-be-powerhouses' of India.

6.3.3 Campus Entrepreneurs Group

As the activities picked up pace, it became essential that entrepreneurial minds scattered in different corners of the campus came together and created something worthy. Therefore, a forum was envisaged to tie together a plethora of activities undertaken by student entrepreneurs in the campus and to enable a healthy exchange of ideas, experiences, insights and challenges in this field.

There are *'Entrepreneurial Bites'* which are mini-conferences over coffee, presided over by the visiting alumni or entrepreneurs. This has helped in establishing the essential connect with the industry. However, some lessons are best learnt in an informal setting, and *'Campus Hangouts'* ensure that our entrepreneurs do not miss those. *'Mentor Connect'* is another initiative by E-Cell, which opens doors to the corporate. All resources, be it monetary, managerial or intellectual, are made accessible, by an active board of mentors, which is here to guide the budding entrepreneurs.

6.3.4 PRIME'83

PRIME'83 is one of the most ambitious undertakings of the E-Cell. The Cell got its first evidence of having grown up when the 1983 IITK batch approached it in 2012 and proposed PRIME'83: Promote Research, Innovation, Mentoring and Entrepreneurship. It was an initiative of our alumni to provide for student entrepreneurs and see them become successful technopreneurs. The SIIC (SIDBI Innovation and Incubation Centre) and the Office of Dean of Resources and Alumni at IIT Kanpur came together to facilitate it. First launched in November 2012, this programme calls for innovative and scalable projects by students who are then mentored and awarded a grant after shortlisting. It includes a corpus of 5 lakhs distributed every year via the PRIME'83 Innovation Awards.

Besides these, E-Cell also organizes TEDx IIT Kanpur, Startup 101 Lectures, ESSENCE and more, to engage students with the expanding entrepreneurship world throughout the year.

6.4 Success Garnered on its Way

So far, more than 30 start-ups across different industries have taken wings. Some of these ventures have reached enviable stature, while others are on their way. There are a few that are still experimenting with ideas, but they too along with the rest, perfectly befit E-Cell's motto 'Startup with your life'.

6.4.1 Transversal e Networks (TeN), One of the Entries for Megabucks

Established in January 2002, TeN has emerged as one of the India's leading e-learning solutions provider with 100+ client installations in India and abroad. It provides key solutions in the domains of learning management, corporate training and online assessment. The technologies and software solutions developed by TeN enable organizations to create a fully web-based environment for constant learning, training and knowledge acquisition and sharing, even if they are geographically distributed. TeN's solutions are full-powered, robust, customizable and scalable. It is now, a product division of UST Global, a California, USA-based IT services company with more than 7500 employees. TeN maintains a strong partnership with a host of prestigious system integrators and marketing partners such as Tata Interactive Systems, HCL Technologies, Xchanging, TCGME, Ecole Solutions and Accel Frontline.

6.4.2 Help Us Green, the Winner of the UpStart Event at E-Summit '15

The venture is based out of Kanpur, a tier-two city of India. It focuses on producing handmade 'flower-cycled' products from holy waste to save the rivers of the country. The team collect the flowers directly from the place of worship and recycle them to produce a wide range of products that include biofertilizers, incense sticks and bathing bars. Underprivileged women living around the area are employed to collect the flowers.

6.4.3 Buizhome, One of the Participants at E-Summit '15

It is an online platform where one can find the best businesses to get his/her work done in a hassle-free fashion. Their vision is to use technology to bring local businesses near you, thereby saving your time. They plan to reduce the waiting time by enabling users to book appointments with the local businesses. This venture too is based out of Kanpur.

6.4.4 Adurcup

Adurcup is an automated hyperlocal advertising tool. It has an offline ad network aimed at disrupting the advertising industry with practices which the users want. Currently, operational in Delhi NCR, the company reaches out to more than 25,000 consumers per day across Delhi NCR and more than 3,000 consumers per day on trains. The inventory comprises of more than 200+ restaurants and cafes in Delhi NCR, top corporate houses in Noida and Gurgaon, travellers on trains and so on. This inventory is growing at

50% month on month, and by the end of the year, they estimate that they would be reaching out to more than 100,000 consumers on a daily basis.

6.4.5 FoodMonk

It is not just a simple restaurant directory; it is an easy to use, highly intuitive online food ordering platform that serves your taste buds as per your choice and convenience. Considered as one of the popular start-ups in the campus, FoodMonk has already tied up with restaurants across Kanpur and Kota and found ready consumers. Now, it is leveraging technology to ensure that people in low-connectivity areas can also place their orders.

The myriad nature of these ventures only posit towards the fact that IITK student community has come of age. Shedding its fears and inhibitions behind, it is ready to take on the world and bring a revolution in the entrepreneurial ecosystem of the country.

6.5 Taking Leaps

E-Cell is taking the entrepreneurial ecosystem of the institute to the next level with Deferred Placement Programme and a comprehensive Student Entrepreneurship Policy (SEP). Deferred placements shall better equip students to take risks and start their own ventures. The SEP will further empower student entrepreneurs by making the environment more conducive for starting up. This policy shall legalize running start-ups as students, with appropriate regulations. It shall open new possibilities for collaboration and commercialization of the technology innovations by faculty and students.

With 15 student start-ups currently active in campus, and several alumni start-ups in constant touch, E-Cell has largely succeeded in becoming the one-stop solution to all entrepreneurial issues: inspiration, promotion, guidance, mentorship and a link to the real playground. It strives to reach out to the 'IITK junta' and convey that the idea of entrepreneurship applies as much in politics, religion, society and arts, as it does in business. It is working towards building an inspired GenNext who has an inbuilt urge to innovate, take risks, shoulder social responsibility and indulge in creative execution of their vibrant ideas.

The Role of Experimentation in Fostering Innovation

7

Bishakh Bhattacharya

Abstract

The central workshop at IIT Kanpur was established in 1966 as a core facility to help students and faculty with the fabrication of mechanical components. However, in the new innovation centric model, the vertical alignment got readjusted into a horizontal framework. This article outlines the existing facilities of the central workshop, discusses the potential of such facilities to initiate the culture of 'total-engineering' and finally identifies a few projections for its future growth and development.

Keywords

Bauhaus philosophy • Technological innovation system • Tinkering space • New product development

7.1 Introduction

In the period of 1919–1933, a group of architects under the leadership of Walter Gropius developed a novel construction school in Germany. This school, popularly known as 'Bauhaus' [1], had propagated a culture of 'total-art' in Germany where all forms of art were taught and practiced together so that complete works of art could be created. The Bauhaus philosophy played an important role in channelizing the traditional architects into the mainstream of product design which was demanded by the rise in the industrially developed modern society. Such was the impact of the school that even though the Nazis closed it subsequently in Germany, the Bauhaus culture spread all over Europe and the USA in the next few decades. The Bauhaus philosophy essentially wanted to harmonize the craftsmanship with modern technology by teaching a balance of form and function for which a unified approach was necessary. Towards the end of the twentieth century, a new wave of development started to take place which emphasized on the integration of technology with market force by a seamless participation of user, builder/technologist, and entrepreneurs. A natural common platform for this activity would be a facility where manufacturers work jointly with the product architects who are ambitious to develop new products. Thus, in this new milieu, the functioning of a central workshop is envisaged to provide a similar platform which the Bauhaus construction school had contributed in the beginning of the last century.

The Central Workshop at IIT Kanpur was established as a core facility to enable students and faculty members with the fabrication of mechanical components. The basic facilities of machining consisting of turning, milling, planing, grinding, abrasive cutting of hard materials, and grinding of cutting tools were developed for this purpose. Nevertheless, as the demand from the growing number of students and faculty increased, miniscule copies of the same model gradually started emerging at different departments/laboratories in a more distributed mode. However, the role of central workshop and its miniature models remained confined mostly to serve the pedagogical needs. Down the years as the emphasis shifted towards creating a technology innovation ecosystem at the institute, the central workshop's role had to be redefined. Currently, the new role is envisaged towards spearheading a culture of complete product development. This involves the development of a creative arena called

B. Bhattacharya (✉)
Department of Mechanical Engineering, IIT Kanpur, Kanpur, India
e-mail: bishakh@iitk.ac.in

B.V. Phani and S. Khandekar (eds.), *Innovation, Incubation and Entrepreneurship*, IITK Directions 1, DOI 10.1007/978-981-10-3334-6_7

'tinkering laboratory', where students can learn the 'tricks' of building a product on their own in the form of making a 'quick and dirty' prototype, modify their product with the help of rapid prototyping facilities, and, subsequently, build up a more professional version with the help of specialized machines.

7.2 Facilities at Central Workshop

As stated earlier, the technology platform in the central workshop was earlier focused on pedagogical training. Hence, the entire workshop was organized functionally into four broad groups: machining, fitting, precision finishing, and painting. However, in the new innovation centric model, such vertical alignment has been readjusted into a horizontal framework. In this framework, machines are reorganized based on the stages of product development. Broadly classifying, any new product development goes through four stages: **Concept Generation**, **Business Analysis and Marketability Tests**, **Embodiment Design and Manufacturing**, **Product Ramping and User Review** [2]. The role of workshop is directly important in two phases: concept testing and screening by model development (Phase 1) and in the phase of embodiment design and manufacturing (Phase 3). It is envisaged that one needs four groups of machines in Phases 1 and 3. In Phase 1, general purpose machines are required. In Phase 3, more sophisticated and heavy-duty machines are required in the first stage, and in the second stage, complex machining is required for taking care of form and ergonomic issues. This is achieved with the use of Special Purpose Machines and Machines for Complex systems. The concept is further elaborated with examples in the following section.

7.2.1 General Purpose Machines

These are the machines which are ideal for tinkering with new product development. Figure 7.1 shows the image of a 'DO ALL' machine, one of the first purchased from the company with the above name. It is basically a power bandsaw which uses a continuous band of metal blade having teeth along one edge to cut various work pieces. The horizontal bandsaw can be used to cut wood and metals such as structural tools at a high speed. Similarly, shapers are used for flat surface shaping, while milling machines are used for surface profiling and common drills for making various types of holes in a structure. It is important to note that students with little training can easily handle these machines on their own. Hence, such machines can be made available to the students for making quick and rough models.

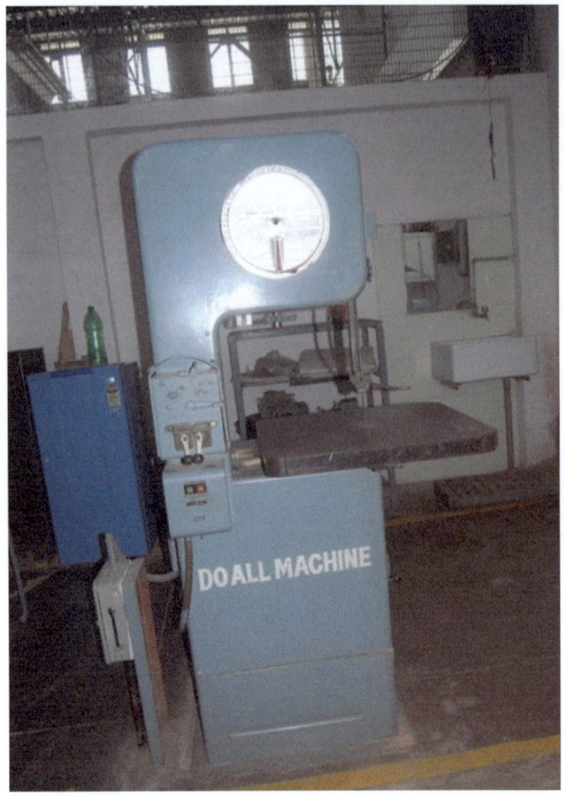

Fig. 7.1 A horizontal band passing machine. [Size of machine: 1300 × 870 × 2135 mm, cutting height: 100 × 300 mm, speed: 0–1450 rpm, blade size: 3760 × 27 × 0.9 mm, generally used to cut flat and thick block of material with manual feeding]

7.2.2 Special Purpose Machines

Often, the form or functioning of a product demands special machining process. Two such examples are provided here as follows: (a) to work on axisymmetric work pieces such as tubes, cylinders, or barrels, general purpose lathe is used frequently; (b) to incise intricate design on a product, engraving machines are required. Figure 7.2 shows a typical lathe machine. It is multifunctional such that the same machine can be used for turning, facing, parting, boring, drilling and knurling. Even though lathe is one of the most ancient machines developed by mankind, it has advanced so much over the last century that specially trained technicians are generally required to use it effectively. Figure 7.3 shows another special purpose machine.

Panto mill-cum-engraving machine is based on parallelogram mechanical linkage system such that the movement of a tracer of an image can generate identical movements in an output engraver. Thus, for a line drawing traced by the input, an identical, enlarged, or miniaturized copy is produced by the output. This has versatile uses in duplication such as sculpting, minting, engraving and milling.

Fig. 7.2 A multipurpose delta lathe machine. [Make: rockwell, swing over cross slide: 150 mm, swing over saddle: 245 mm, distance between centre: 800 mm, spindle speed = 45–1550 rev/min]

Fig. 7.3 A panto mill-cum-engraving machine. [Make: panto mill gorton, size of work table: 8" X 18", cross travel = 8" & vertical travel = 10", pantograph ratio = 1:2 & 1:16]

7.2.3 Machines for Mass-Scale Manufacturing

In any professional manufacturing environment, one needs heavy-duty machines to enhance productivity and maintain quality. These types of machines, however, need employment of full-time professionals. Figure 7.4 shows one such heavy-duty lathe machine. Such machines have broad range of capabilities such as handling jobs of varied size as well as multifunctional features which enable them to be used for milling, shaping and grinding. Another important feature of this type of lathe is the availability of a wide range of rotational speed, thus enabling a flexible manufacturing process. Both special purpose machines and machines for mass manufacturing demand technically trained manpower. The availability of such skilled manpower, however, is difficult in an academic institute set-up. One strategy of circumventing this problem is to build up a viable teaching–apprenticeship relationship with the trade-based technical training institutes such as ITI. For example, IIT Kanpur has started to use this route successfully by signing an MOU for manpower training with a local ITI institute named JK Training Centre at Kanpur.

7.2.4 Machines for Complex Systems

With the rapid advancement of product development today, there are two major manufacturing challenges for the new product development. One is the continuous reduction of lead time, and the other is integrating many functionalities/features into a single product. Complex machining processes are required to meet these challenges. Additive manufacturing process has enormous potential to meet this challenge. Both designers and product incubators today have received competitive advantages due to the scope of additive machines such as 3D printers and reverse engineering using 3D scanners. However, this facility needs highly educated manpower. Hence, close association with the manufacturing laboratories/research establishments is crucial for successful operation of the system.

7.3 Workshop as a Medium of Technological Innovation System

A technological innovation system is defined as an ecosystem that enables technology to be transferred in a societal context rather than keeping the knowledge confined in its own domain. Thus, a technology innovation system encourages individual players to assimilate the advantages of a particular technology and transfers them into a new product. The new product should eventually get exposed to the market force and undergo incubation for incorporating 'new genes' that can help it to survive in a competitive domain. The four significant factors in this direction are as follows: knowledge development, knowledge diffusion, entrepreneurial activities and guidance of the search [3, 4].

Fig. 7.4 A heavy-duty lathe machine. [Model: 18SL, Make: redman churchill, centre lathe with straight bed: 11'-6" long, spindle speed: 15–780 rpm]

While knowledge development and diffusion can take place within a technical domain, for the last two activities, it is imperative that the workshop gets linked with the start-ups and facilitates marketing of the products. Thus, the survival of the workshop today is hinged on its capability to get integrated into a four-dimensional ecosystem: joint programmes with the trade training centres, knowledge development at the graduate and post-graduate level, facilitating entrepreneurs and maintaining a strong presence in the in-house research and development. This will not only usher in a harmony between the builders, sellers and the users of the technology but also help to develop a new 'Bauhaus model' of innovative product development which is critical for the success of our manufacturing industry.

Acknowledgements The author wishes to acknowledge Mr. Sharma and Mr. Rajesh Srivastav of the Central Workshop for their inputs of images and information procurement related to the machines of IIT Kanpur Workshop.

References

1. Wood P (1999) The challenge of the avant-garde. Yale University Press, New Haven. ISBN 0-300-07762-9
2. http://www.innovationexcellence.com/blog/2013/05/27/8-step-process-perfects-new-product-development/
3. Smits REHM (2002) Innovation studies in the 21st century. Technol Forecast Soc Chang 69:861–883
4. Carlsson B, Stankiewicz R (1991) On the nature, function, and composition of technological systems. J Evol Econ 1:93–118

Withering to Weathering—A Farmers' Friend

8

Sonu Agrawal

Abstract

Weather Insurance and Derivatives is a case of multiple technology disciplines coming together to offer a unique solution for farmers to protect them against vagaries of weather and climate change. Credited with pioneering weather insurance in India, Weather Risk Management Services founded in 2004 has emerged as a global climate risk management company, offering a wide range of products and services combining big data, analytics, smart devices and financial services. In its 12-year history of credibility with key market players, the company is in a position to address an INR 10,000 Cr. market. This article highlights the innovative solutions developed by the company, and the humongous impact created by them.

Keywords

Climate risk • Weather index insurance • Big data analytics

8.1 Introduction

Natural disasters are a recurrent phenomenon in India. Out of 36 States/Union Territories, 25 are prone to floods, cyclones, droughts, earthquakes and other natural calamities. As per the National Flood Commission, about 40 million hectares of land is prone to floods. About 18.6 million hectares of land area and 3.7 million hectares of crop area get affected by floods on an annual average basis. Besides, 68% of the area is prone to droughts and 56% to the seismic activities. Some other disasters that hit the country from time to time include landslides, hailstorms and avalanches. The susceptibility to disasters is compounded by frequent occurrences of manmade disasters such as fire and epidemics.

Every time a disaster strikes, it is the poor of the country who is most affected. There is a huge loss of human life, crops, livestock and other valuable assets. Furthermore, its impact is covariate and severe so much so that the various means usually adopted by the poor to manage these risks such as inter-family borrowings and sale of assets fall inadequate. As a result, the poor is pushed towards resorting extreme strategies such as external borrowings and migration. It has also been clearly demonstrated by the economic development literature that the poor end up paying for more risks when the risk transfer instruments are not available to them.

Innovative insurance mechanisms (weather index insurance) can play a vital role in addressing the several risks faced by the poor and also help in creating synergy among insurance, savings and credit markets. This synergy cannot be attained by merely focusing on one financial service such as increasing credit availability. The new insurance mechanisms transfer the risks associated with weather changes and natural calamities out of rural areas of lower income countries, thus lowering the risk exposure of poor households and giving them access to broader financial services at better terms.

S. Agrawal (✉)
Weather Risk Management Services Ltd., New Delhi, India
e-mail: Sonu.agrawal@weather-risk.com

B.V. Phani and S. Khandekar (eds.), *Innovation, Incubation and Entrepreneurship*,
IITK Directions 1, DOI 10.1007/978-981-10-3334-6_8

8.2 Weather Risk Management Services

Climate risk represents the daily reality of hundreds of millions of farmers and directly impacts the huge ecosystem of enterprises that engages with them in the global agricultural supply chain with implications running into billions of dollars. With a vision of providing security against climate change to all in the world, Weather Risk Management Services Pvt. Ltd. was set up by Mr. Sonu Agrawal in 2004. With ICICI Lombard as its first client, the company today works with every insurance and enterprise player serious about transforming agriculture in India creating direct change for over 1 million farmers. The multidisciplinary technology expertise of the 100+ team, integrated capabilities in Big Data, Manufacturing, Technology and Underwriting, and high order R&D built-in ecosystem have powered diversified risk products. It is pursuing active collaborations with research institutes on Sensing Systems, Geo-forecasting, Big Data algorithms, Energy Efficiency to evolve products.

8.3 Products and Services

8.3.1 Climate Finance

8.3.1.1 Weather Secure

It is an advanced and comprehensive weather risk management platform for global insurers, reinsurers and agricultural businesses. It provides Crop Insurance Portfolio Analytics and Monitoring tools. The **WeatherSecure accuyield®** module is a robust and mature crop yield forecasting facility with the capability to use seasonal forecasts and find dependencies on weather–soil–irrigation–technology. It includes an advanced disease management and forecasting module that can draw on insights for over 100 pests and diseases. It provides the following:

- 30×30 km^2 weekly weather forecasts,
- Regional monthly and seasonal weather forecasts,
- Yield data at 30×30 km^2,
- Satellite-based vegetative indices at 50 km^2,
- Comprehensive disease, flood and drought alerts.

WeatherSecure HDM® is a robust historical data manager that contains historical weather parametric data for each and every state of India. It uses dedicated modules for data cleaning and comparison with nearby locations.

WeatherSecure Portfolio Builder® can build multiple portfolios based on weights and sales scenarios and can seamlessly execute VAR Distribution fitting and de-trending on Portfolio payoffs. Portfolio is automatically updated for any change in individual term sheets. Tools allow easy additions and deletions in the portfolio and can display both basic and advanced Portfolio Statistics.

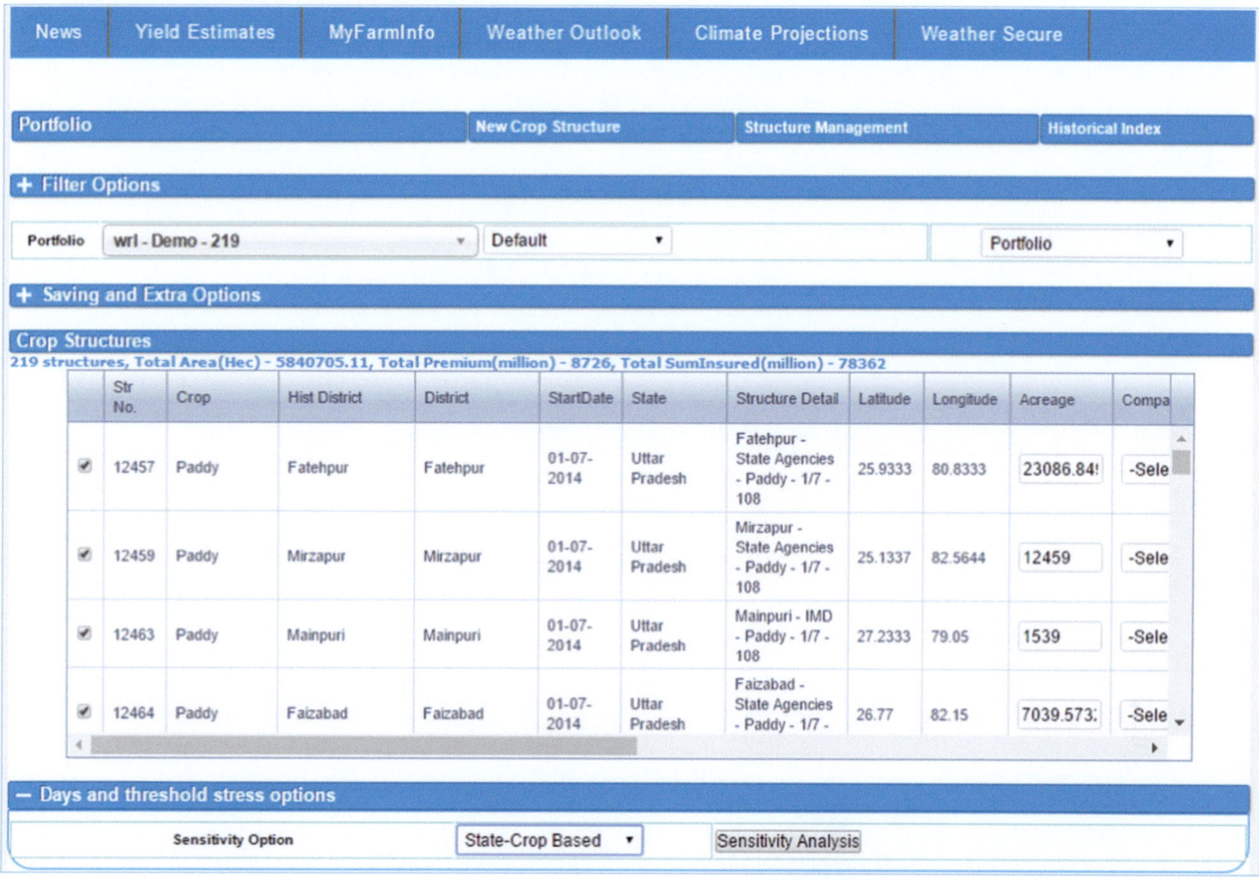

Snapshots of Weather Secure Portfolio Section

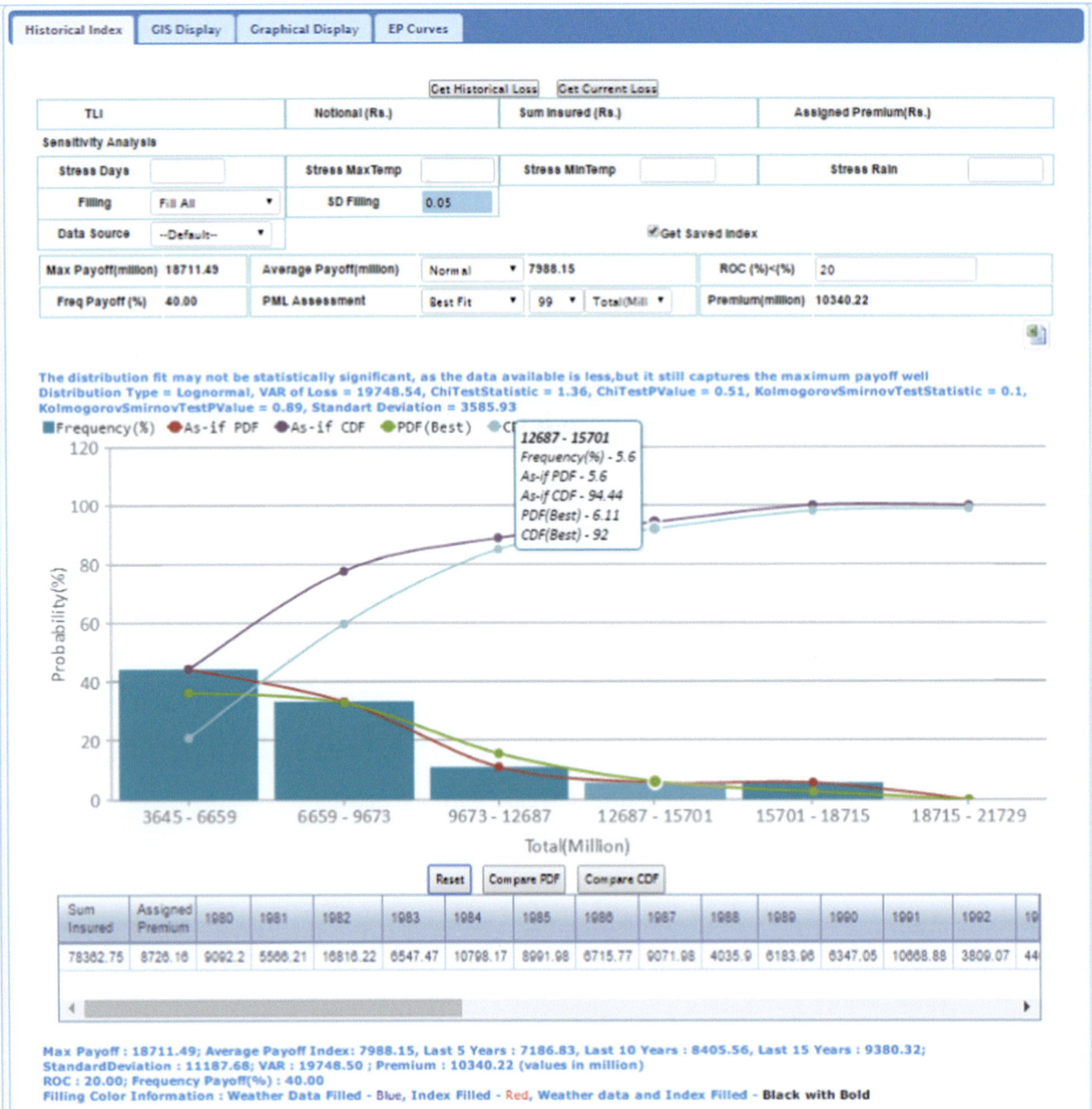

Snapshots of Weather Secure Portfolio Section

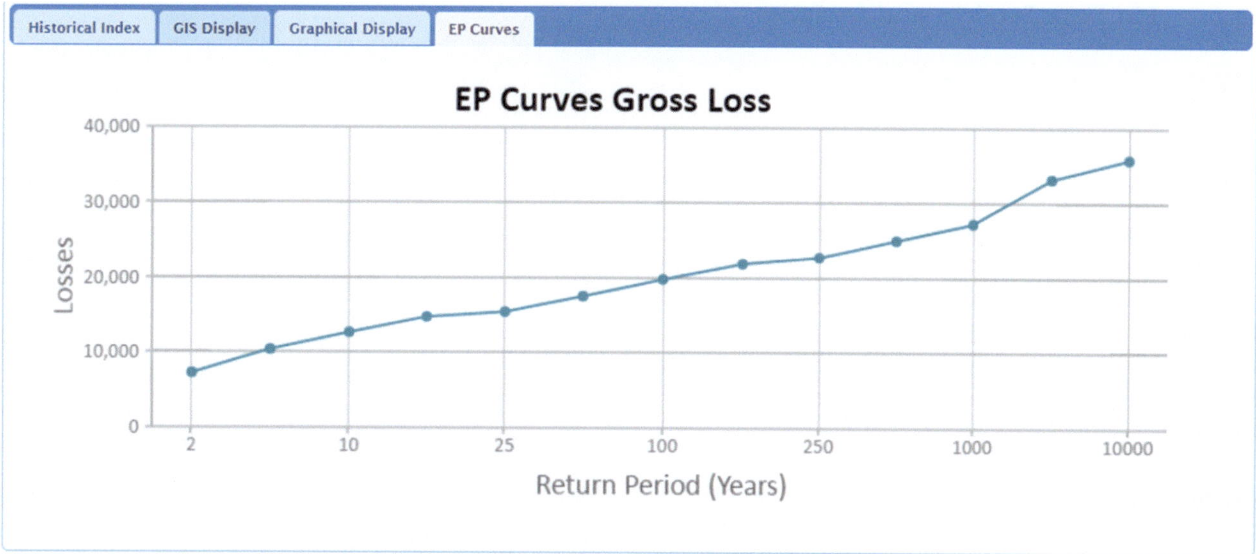

WeatherSecure Claims Manager® is a real-time claims monitoring and assessment solution for both Weather and Crop Insurance. It has dedicated modules that allow seamless GIS mapping of claims. The platform allows Real-Time Monitoring if data is uploaded dynamically.

CASE STUDY 1

Cholamandalam is a leading Indian general insurance company. In 2011, Weather Risk was hired to help structure Weather and Yield Index-based Crop Insurance Products.

Business case	What the company did	Impact
Cholamandalam wanted to establish a robust Crop Insurance Products portfolio	• Helped to structure Weather and Yield Index-based products as well as advised Cholamandalam to manage its Crop Insurance portfolio • Helped to build necessary banking channels for sales and distribution • Installed Weather Stations and conducted claim settling audits	• Cholamandalam generated a premium income of US $20 m in FY 13–14 • The portfolio covers 20,000 farmers in Bihar and Rajasthan • WRL continues to help Cholamandalam target US $40 m in 2014–2015

CASE STUDY 2

Since 2007, WRMS has helped **ICICI Lombard** implement a joint nationwide weather and agriculture Insurance market making and strategic development exercise.

Business case	What the company did	Impact
Large scale implementation of subsidized weather insurance program	• Embarked on an exercise to remove technological, policy level and reinsurance capacity bottlenecks • Developed low-cost weather stations, unmanned aerial vehicles and automated yield measurement instruments to facilitate claim settlement • Pursued the Ministry of agriculture for subsidy support	• The Ministry kindly consented in 2007 and granted support to Weather Insurance Products • In Rabi 2009–2010 season, WRL covered close to 4000 farmers for 10,000 acres in just two districts of West Bengal

8.3.2 Big Data Services and Analytics

8.3.2.1 Precision Farming Solutions

Accufarm® is a whole farm management approach optimizing returns on inputs while preserving scarce and valuable resources. It is not only a solution for yield mapping and variable rate fertilizer, but also a window to the entire production function and the management function of the farm which allows for improved economic analyses. It uses object-based hierarchical image analysis to classify plot imagery from UAVs and Satellites measured concurrently on the ground using standard rangeland monitoring procedures.

Benefits of using accufarm® Enterprise

Site feasibility and project planning	Soil and PH testing; variability analysis; crop modelling using local climate, ground water, soil information; remedial measures; value maximization by price and quality matrix; credit and project finance; risk mitigation through insurance
Social impact assessment and management	Women participation; opportunities for livelihood enhancement; labour engagement and relationships; agro-forestry practices
Technical assistance in commercial production	Soil testing and nutrient recommendations in a precision farming framework, pre-yield assessment through GIS, field surveys and agro-meteorological models **Input output optimization**: Crop scheduling by weather and climate; soil reclamation; water reclamation; forecasts and pest management support; weather and seasonal forecasts for farmers
ICT integration	Real-time farm monitoring and management through web platforms Outreach to farmers through mobile platforms (voice, text, videos)
Output management	Food safety; crop output assurance; facilitation of primary and secondary market linkage post harvest management and packaging interventions

CASE STUDY 3

Horticulture Risk: Sheikpura Polyhouse/Karnal, Haryana

In a live project underway, Weather Risk is engaged with a group of farmers in Haryana invested in polyhouses covering multiple acres of land to help recover investments against pests and disease. The accufarm® methodology involved detailed profiling of the ecosystem—the polyhouse structure, weather, soil and PoP, as well as a plan for soil treatment and pest metering across the farm supported by constant vigilance using pheromone traps and mobile cameras. It used disease forecasting models along with WRF-generated forecasts at 9 km^2 providing for immediate control in case of disease forecasts or observation. WRMS could guarantee compensation for yield loss example: <30 tons/acre for cucumber.

CASE STUDY 4

The company has helped transform **Sowing Failure Insurance** for farmers covered by Mahyco by providing comprehensive risk management with a first of its kind product in India.

Business case	What the company did	Impact
Innovation for sowing Failure Insurance	• Launched innovative sowing failure weather insurance, which was offered free of cost along with the seed packet • Premium for the insurance product paid by Mahyco and insured by HDFC Chubb • Designed a very effective marketing campaign using radio and newspapers	• Helps inputs providers to increase sales of new varieties • Farmers find the comprehensive risk management very useful • This insurance has become a regular marketing feature for seed and input providers

8.3.2.2 Smart Grid

Accunergy® is a path-breaking, load-forecasting, control and estimation technology for the power industry. It is an advanced smart grid solution designed to assist distributors to conduct better load forecasting and helps transform current approaches used in estimating transmission losses in profitable zones. In non-profitable zones, power companies can adopt this pioneering technology for control, managing pilferage and transmission losses at a very attractive cost to benefit advantage. Matrix helps in advance estimation of supply gaps and can directly assist strategies to make temporary arrangements to mitigate supply gaps. It is also a robust planning tool that can estimate monthly demand and surges over 24–48 h.

The product has been developed in technical collaboration with the Rensselaer Polytechnic Institute and Centre Suisse d' Electronique et de Microtechnique.

CASE STUDY 5

For two of India's largest power distribution majors in Delhi and Mumbai, BSES and Reliance Infrastructure, WRMS is helping derive significant savings, planning efficiency and the ability to look ahead for weather impact on load and demand.

Business case	What the company did	Impact
Indian power distribution major: Provides electricity to 6 million households in Delhi and Mumbai	• Provided 7-day FORWARD weather forecast for every 15 min with intraday revisions and weekly weather trends • 24 h FORWARD load forecast using weather forecasts, data, social media trends, regular consumer surveys • Energy audits in large hotels and buildings	• Significant savings for the client as it is able to purchase power in advance and also manage customer expectations especially in cases when the power consumption drops due to sudden fall in temperature

8.3.3 Smart Devices and Technology

8.3.3.1 Automated Weather Stations

Ingen Automated Weather Station (AWS) provides continuous observations on basic weather parameters such as air temperature, humidity, soil moisture, dew point, sunlight intensity and rainfall. These parameters are the most crucial for monitoring agricultural production. It is easy and quick to install. The AWS software generates weather message automatically every 30 s for 3 h: the user has to simply select the interval. The messages are stored on the PC hard disk and if user can edit the weather messages manually using the AWS SYNOP Editing software, he/she can view the sensor data in real time. The AWS can be provided with an optional memory unit for data storage. The built-in algorithms test each measurement to ensure data quality. For each parameter, tests are carried out on the minimum, maximum and step limits and the different parameters are crosschecked. A built-in testing system also continuously checks the hardware and reports immediately if a fault occurs. The systems have been designed and manufactured in IMD and ISO certified facility.

Ingen Automated Weather Stations

Largest Private Network: AWS Pan India Network

8.3.3.2 Vehicle Tracking and Fuel Monitoring Systems

The company has indigenously developed and installed a wide range of tracking systems such as iTrack-P10GPS Bike Tracker, iTrack-M8 VTS device, iGDL-L80 GPS Data Logger that:

- Monitor and Save fuel.
- Prevent theft and pilferage.
- Send SMS alert and live tracking on mobiles.
- Track your vehicle anywhere in the world live.
- Save transportation cost.

8.3.3.3 Smart Energy

The automatic irrigation controllers enable farmers to regulate irrigation supply according to the weather. The irrigation system can be switched on or off remotely through a simple SMS. One can set the system in auto irrigation mode based on pre-defined weather conditions.

The PV string monitoring unit allows owners and integrators of PV systems to monitor the performance of each string or subarray by providing direct current readings. It can monitor 8, 16 or 24 strings using shunt and provides a Modbus RS-485 and X-bee output.

8.4 A Global Footprint

Working painstakingly over a decade in a tough regulatory regime, Weather Risk has finally succeeded in registering profits. The company is in a position to address an INR 10,000 Cr. market in the next 4 years. It has gradually expanded its global footprint, and discussions are underway in ASEAN, Africa, Europe and China

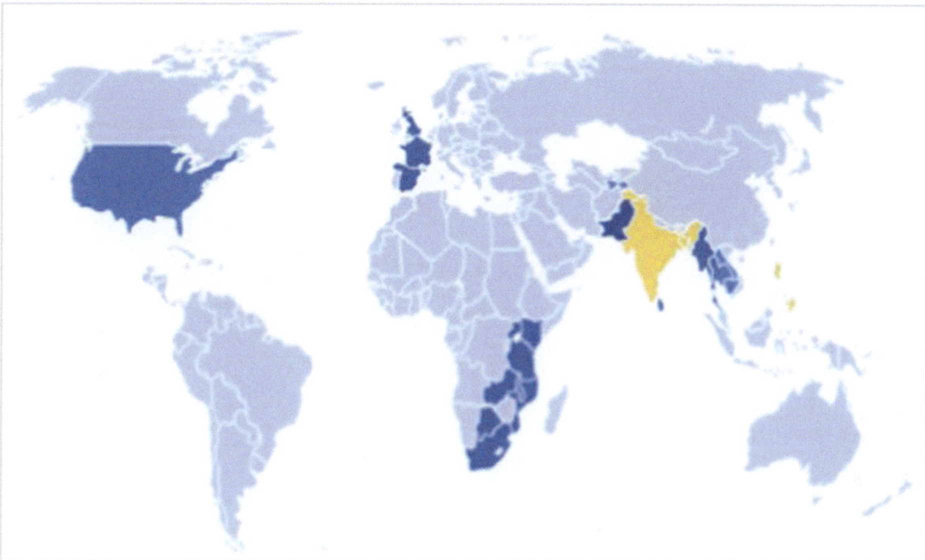

WRMS Global Footprint

Besides several successful case studies mentioned above, it has created a huge social impact through its work.

- Served 1 million farmers across India transforming lives and agriculture.
- Bailed 500,000 farmers struck by natural disaster.
- Catalyst for the growth of Crop Insurance in India from INR 300 Cr to 5000 Cr in 10 years creating employment for ~10,000 people.

8.5 The Road Ahead

Weather Risk Management Services has emerged as a highly profitable venture and is all set to scale new heights. The company's founders and the dynamic team have defined highly strategized goals for future. It is targeting a turnover of INR 100 Cr. in next 4 years and expanding its share from 1 to 5% of 500 Cr. to 3–10% of INR 10,000 Cr. Premium.

In addition, it also envisages to bring a deep social impact through its services. It is determined to (a) Cover 10 million farmers under crop insurance in the less developed states of UP, Rajasthan, MP, Jharkhand, Orissa, Bengal and Bihar; (b) Install 5000 Stations and engage 500 Agri Surveyors; (c) ACCUFARM® Services for these regions in collaboration with FPOs/SAFAL/Pepsico/Bayer CropScience; and (d) Direct employment to 2000 people in focus states and state of art online.

Winds of Change

9

Rita Singh

Abstract
A large section of the Indian population resides in rural areas where people struggle hard to make their two ends meet. If not all, at least a few of their problems can be solved by the intervention of technology. Though advancements have been made in the field of science and technology, people on the cutting edge of technology have yet to provide technology solutions that suit the demands of the rural poor. With the vision of bridging this gap and bringing technology to rural areas, the Office of Principal Scientific Advisor has set up Rural Technology Action Groups (RuTAG) at Indian Institutes of Technology (IITs). The main idea behind this initiative is to provide a single platform for various S&T organizations, voluntary organizations, public sector undertakings and corporate houses committed to rural development, and state and central government organizations working in rural sector to connect and work under one umbrella.

Keywords
Rural technologies • Bottom-up development model • Organic growth

9.1 Introduction

The RuTAG centre at IIT Kanpur was set up in November 2013 with the aim of developing and disseminating low-cost affordable technologies that rural people of Uttar Pradesh could easily embrace using local resources, thus promoting self-sufficiency. While inaugurating the centre, Dr. R. Chidambaram, the Principal Scientific Advisor (PSA), Government of India, said, “In the case of rural development, tremendous knowledge with academia has somehow failed to reach the places where a huge demand exists. RuTAG largely focuses on delivering solutions to people who need it the most”. The centre proposed to take up both small and big projects, where problems would be addressed either by making incremental improvements in technologies that already exist, customizing solutions to local conditions or by working on completely new ideas (Fig. 9.1).

RuTAG is a synergizing and catalysing mechanism and not a major funding mechanism. Professor Nalinaksh S. Vyas (ME) heads the RuTAG Centre at IIT Kanpur. The Core Team includes Professor S Sundar K Iyer (EE), Professor J Ram Kumar (ME) and Professor Arun K Sharma (HSS). Under the RuTAG umbrella, individual projects are taken up independently by teams of faculty, staff and students. These projects are approved and monitored by the Project Advisory Committee whose members are from the core group, industry and social sector.

9.2 RuTAG Model of Development

RuTAG employs a *land to lab* model of technology development. In this bottom-up model, the end-user is involved in the early stages of project vision itself. Problems are identified through brainstorming workshops held with community-based organizations (CBOs) and non-governmental organizations (NGOs) working with rural communities in different parts of the state. The result is a

R. Singh (✉)
RuTAG, IIT Kanpur, Kanpur, India
e-mail: ritas@iitk.ac.in

B.V. Phani and S. Khandekar (eds.), *Innovation, Incubation and Entrepreneurship*,
IITK Directions 1, DOI 10.1007/978-981-10-3334-6_9

Fig. 9.1 Inauguration of RuTAG IIT Kanpur

continuous involvement of local people in the technology development process. This ensures that technologies are developed based on the needs of people rather than adopting a "push a technology" strategy.

A critical ingredient of this effort is to build on peoples' knowledge and grow in a manner that rural entrepreneurship gets nurtured. Indian craftsmen, farmers, mechanics and rural communities have continuously showed ingenuity to solve their local problems and exhibited innovativeness in developing new products to suit their needs. These are quite often *jugaads* that involve low cost and often pertain to incremental innovation like process improvement. Injection of appropriate technologies can help rural entrepreneurs upgrade themselves to better designed and scalable technologies using frugal local resources. Recognizing the vast size of informal sector and role played by grass-roots innovators, RuTAG invites these innovators to hold discussions and interactions with the team at IIT Kanpur to develop their product ideas.

In a short span of 2 years, RuTAG has succeeded in building partnerships with several technical institutions including the Indian Council of Agriculture Research, IIT BHU, Indian Institute of Vegetable Research, Varanasi, and UP Textile Training Institute. The domain-specific knowledge of scientists working in these institutes is being sought on specific projects, thereby laying foundations for an organic growth of RuTAG, i.e., linking people and local resources.

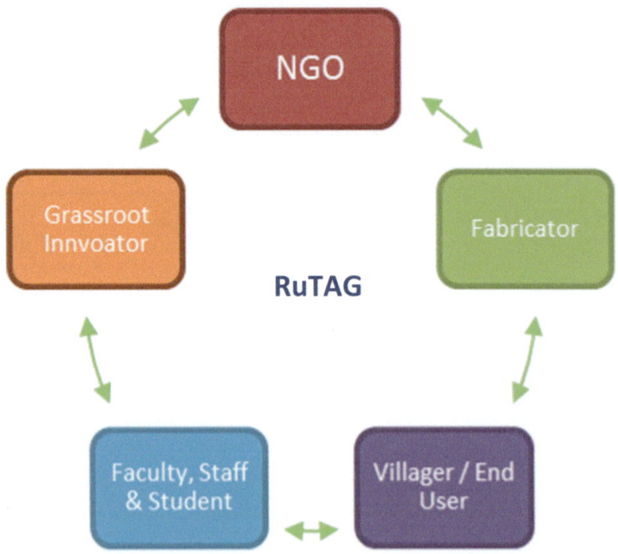

9.2.1 A Platform for Social Entrepreneurs

Though in its nascent stages, RuTAG is striving to develop an ecosystem for students to connect with rural India, work on real-life technology problems and innovate simple solutions using minimal resources. While working on the problems, students are required to make field trips to the areas, interact with village communities in local dialects, absorb

Fig. 9.2 Summer interns demonstrating prototypes to users, July 2015

their knowledge and practices, understand their needs and identify problems and then deliver solutions. These novel experiences not only leave them enriched with ever lasting impressions but also broaden their horizons as they are required to think practically and differently. A few of these students have already started thinking of setting up their own enterprises.

So far, the centre has witnessed phenomenal response with more and more students hailing from Tier II and Tier III cities coming forward to be a part of RuTAG projects. Some work as volunteers and interns, while the others opt them as Under Graduate Projects (UGPs) and thesis topics. The number of summer interns has grown from 4 in 2014 to 18 in 2015. The summer internship program 2015 attracted engineering students from institutes like IIT Gandhinagar, IIT BHU, NIT Hamirpur, NIT Surat and ISM Dhanbad (Fig. 9.2).

9.3 Identification of Problems

Since its inception, RuTAG has held numerous brainstorming sessions and workshops with local organizations and communities which has resulted in the identification of several relevant problems to work on. Since the geographic area of the state of Uttar Pradesh is large, problems are complex and diverse and therefore solutions need to be multiple, local, energy efficient and sustainable. The team is planning to take up a few of these listed problems in the coming year.

Problems	Possible solutions
Vegetable production is on rise in the state, but post-harvest losses are huge. Farmers end up doing distress sale resulting in huge losses	A low-cost cold chamber that can increase the shelf life of vegetables such as tomatoes and spinach
In flood-affected areas, food packets catch fungus quickly. This problem was brought to notice by an NGO from Bahraich, UP	A low-cost dry bag that can keep food fresh for a couple of days
The Indian callipers do not bend more than 90°, thus restricting the mobility of polio-affected children who use them. The imported callipers with more flexibility cost about Rs. 1 lakh	Callipers that bend more thus enabling a child to sit. This would change the lives of lakhs of children by giving them additional mobility. Most of the users will be children from poor families
The sheep wool produced in the region of Sonbhadra is short and coarse and therefore is not used for yarn making	Redesigning the traditional charkha so that it takes staples of size less than 1.5. Such a charkha will help the poor sheep owners of Sonbhadra to produce yarn independently and fetch higher returns
The Banarasi sari weavers use jacquard handlooms in which the design appears on the reverse side. Besides, the loom with large number of hooks (say 400) is heavy to operate	A handloom that can make the design visible on the front side as well and ease the detection of defects promptly. Can there be some way to reduce the shedding load that would reduce drudgery of operation?

9.4 Successfully Delivered Technology Solutions

9.4.1 Horseshoe Making Process

One of the successful technologies coming out of RuTAG is a new process of horseshoe making that enhances the life of a horseshoe from 5–7 days to about 25 days, nearly a fourfold increase. The innovators of this technology are Professor Sandeep Sangal (MSE) and Mr. K Chandrasekhar (Technical Assistant, MSE) who, through an NGO partner Shramik Bharti, worked with 14 horse owners for about a year to come up with a novel technique of improving the metallurgy in such a way that local blacksmith could easily follow. As a result of this technology, horses are saved from the agony of frequent nailing and the cost to a horse owner is reduced. Considering the benefits of this technology, the Khadi and Village Industries Board of Uttar Pradesh is offering finance to shoemakers for adopting the new technology and setting up model units. Brooke's India, a funding agency working on the welfare of horses, is taking this technology to other regions of the country.

Prof. S Sangal, who developed the technology with his team

North India has a large equine population

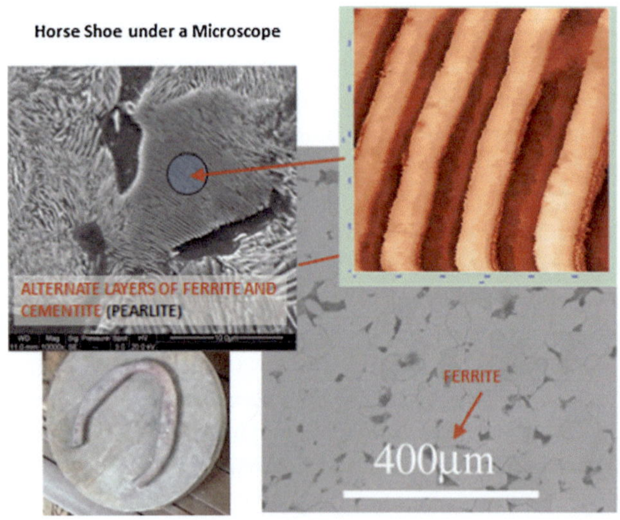

Horseshoe under a microscope

Horseshoe being nailed

Case Study: Bringing smiles on the faces of horse owners

Raju's best companion is Badal, his 8-year-old horse. The morning starts for them by 10:00 am when both go to Transport Nagar for work. On a normal day, they manage to get two rounds, dropping boxes load full of auto parts, in different areas of the city. Badal is strong enough to carry 4–5 quintal load on his back and can run for 5–6 h every day. He is the only source of livelihood that Raju has, who saves around Rs. 250 per day. Because roads today have hard surface, Badal is made to wear horseshoe that not only protects his hoof but also provides good traction and acts as shock absorbent. Guddu is a professional farrier and carefully trims his hoof wall before nailing the shoe in. But still quite often, there is a shoeing mistake and the nail gets into the sensitive part of the hoof. Badal expresses his utter dislike by hitting and pawing his leg. Guddu tells that horses like Badal have to be nailed every 6–8 days because the horseshoe wears off. Frequent shoeing is not good for the health of a horse; it is a financial burden on horse owners. One time nailing costs Rs. 80–100/-, and owner suffers one-day loss of work!

Almost similar problem is faced by all 30,000 odd horse owners who are working in and around Transport Nagar area. A team of researchers from IIT Kanpur led by Prof. Sandeep Sangal and K Chandrasekhar examined the problem closely. With the help of Rana Singh and Rakesh Pande of Shramik Bharti, they worked with the horse owners, farriers and shoemakers of Kanpur region. With the constant effort of the team, the shoemakers have an improved technology that has increased the life of a horseshoe by four times. Guddu farrier says, "The horse owners don't mind paying Rs. 20 extra for this new horseshoe as it lasts for about a month. They demand for IIT wali naal".

Horseshoe being made at a slum in Afeem Kothi, Kanpur

While the technique being offered to shoemakers is simple to follow, it is a novel development in the area of materials engineering. It took almost a year to develop and field test this technique during which more than 300 horseshoes were made with the help of two shoemakers located at a slum in AfeemKothi and shoeing was done by 3–4 skilled farriers. Regular field measurements were made on the horseshoes to observe how it wears and laboratory tests were done to see the microstructure of the material.

This has been a gratifying experience for researchers to step out of their sophisticated laboratories and work in tandem with local NGOs, shoemakers, horse owners and farriers.

9.4.2 Amla Pricking Machine

The second successful innovation coming out of RuTAG is in the food processing sector. A RuTAG team under the leadership of Professor J. Ramkumar has focused its attention on the food processing sector that offers a huge potential to a state like Uttar Pradesh. At present, less than 2% of the fruits and vegetables produced are getting processed. For the food processing sector to truly emerge as an engine of economic growth, there are several impediments that need to be removed. The processors need to mechanize some of the operations that result in low processing efficiency and poor-quality products. Induction of technology is critical for quality enhancement and the processors are beginning to look at it for capacity expansion and upgradation.

A quick look at the food processing machines in the country shows that while most equipment is imported, they cater to large-size operations. Suitable food processing machinery for cottage and small scale operations is not available. The rural landscape of this state has more processors in cottage sector and SSI, which also suffers frequent power outages. RuTAG started working on machinery that would be much more affordable, preferably with an option of manual operation and ease of maintenance. This would ensure that villagers do not depend on expensive imported spares and can easily get it repaired by local mechanics. Below is the case study of amla pricking machine developed by a team of students led by Profs. J. Ramkumar and Shantanu Agarwal and supported by Anil Jha, Rakesh Thaplial and Om Prakash (Technical staff, IIT Kanpur). The machine has been patented and the fabrication work has been handed over to an Ambala-based food machinery manufacturer.

Dr. J Ramkumar who lead the team

A prototype of amla pricking machine

Case Study—Development of a Cycle-operated Amla Pricking Machine

From 2014, RuTAG IIT Kanpur started working in a cluster of villages where amla (*Phyllanthus emblica*) is traditionally cultivated and processed. The cluster is located in Pratapgarh, one of the most backward districts of Uttar Pradesh. The best-quality amla fruits in the country come from the twin villages of Gonde and Chilbila that also house a number of nurseries exporting amla saplings. Hundreds of cottage units processing amla have grown along the sides and typically produce murabba and laddoos. Till about a decade back, this could not be termed as an industry. It was only towards the end of 1990s that a few local entrepreneurs jumped into the fray and began developing a range of amla products that market immediately lapped up. Today, the cluster houses 8–10 processing units with their own market brands (Indica, Pushpanjali, Maya, Vatika, Dolly to name a few), and about 100 smaller units with cottage-level operations. The processed amla sales soared rapidly in last few years supported by interesting mix of product range innovated locally.

A team of RuTAG IIT Kanpur covered this 250 km bumpy ride by road and reached Chilbila. It was immediately noticed that this small village lacks basic amenities and also struggles with erratic power supply. The processors realized that for industry to grow they need to upgrade their production, improve quality and enhance productivity. Since processing operations are handled manually, shortage of labour and space are the limiting factors. The food processing machinery and technological solutions available in the market are for large-size operations. The bulk of market, comprising of small-size processors doing 1–2 quintal of food processing per day, is not catered by them. There is a strong demand for small manually operated machines to prick amla, grate amla, cut burfi, make ladoos, sort amla fruit by size and deseed amla for candies.

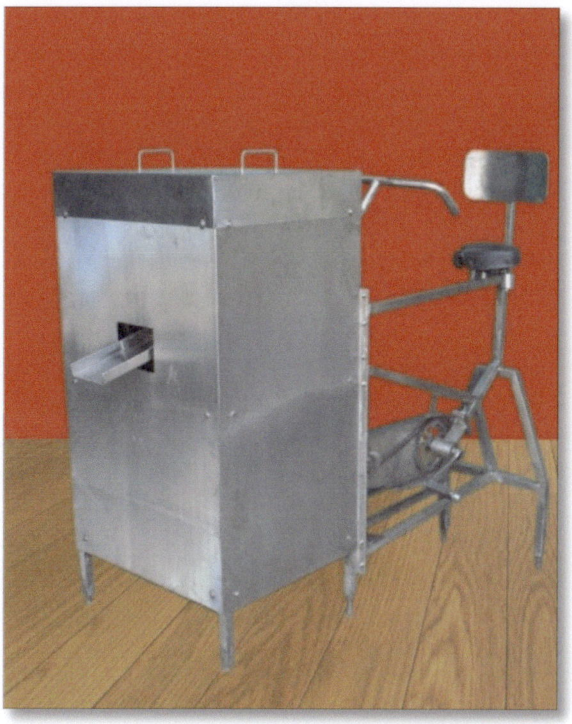

Commercialversion of the machine fabricated by NuTECH, Ambala

As soon as the problem got identified, RuTAG floated several projects that were taken up by few interested students either as summer internships or as Under Graduate Projects. Work began on cycle-operated amla pricking machine by a team of three B.Tech students under the guidance of Prof. Ram Kumar and support of staff members. They succeeded in making a functional prototype which was sent to Pratapgarh for testing by end-users. The machine enhanced productivity from 3 kg (manually) to 40 kg per hour. Now, the end-user feedback has been incorporated and machine design has been given to a food processing machinery manufacturer in Ambala who is producing the final version of the machine. The five developers of the machine have earned a patent for innovation in design. The processors of Pratapgarh are eagerly waiting for the machine to arrive in the market. The fabricator is enthusiastic about the sales of machines as he has already received several queries from potential buyers.

9.5 The Way Forward

RuTAG IIT Kanpur is still in its early years and taking small steps. It is gradually spreading its wings. From one NGO partner in the first year to 15 at the moment, it has expanded its base. The area of action, in the first year, was limited to Kanpur and Pratapgarh; and now it has expanded to Varanasi and Sonbhadra region. It is planning to expand and bring in more NGOs and CBOs working in far flung regions of Uttar Pradesh.

The strength of RuTAG lies in its members comprising faculty, staff and students. From six faculty members associated with RuTAG in the first year, the number has now surpassed 20. Technical staff has volunteered their time after they complete their laboratory work and have willingly given time over the weekends. Students are enjoying their association and are planning to have an active RuTAG Club this year. Interestingly students of humanities, management and sciences are coming forward and exploring ways to connect to rural India. The RuTAG team has also filed for two patents.

RuTAG is a small effort to reach out to those at the bottom of the pyramid through innovative and frugal technologies. This effort can be scaled up only with the support of the whole community. This will happen when working for rural India becomes a passion for everyone, when local geniuses and craftsmen get acknowledged for their ingenuity, when rural entrepreneurs walk into IITs demanding for a technological solution to their problems and when the technologies get evaluated by the social impact they create.

Index

B.V. Phani and S. Khandekar (eds.), *Innovation, Incubation and Entrepreneurship*,
IITK Directions 1, DOI 10.1007/978-981-10-3334-6

MIX
Papier aus verantwortungsvollen Quellen
Paper from responsible sources
FSC® C105338

If you have any concerns about our products,
you can contact us on
ProductSafety@springernature.com

In case Publisher is established outside the EU,
the EU authorized representative is:
Springer Nature Customer Service Center GmbH
Europaplatz 3, 69115 Heidelberg, Germany

Printed by Libri Plureos GmbH
in Hamburg, Germany